The State of Physics
at the End of the 20th Century

In Honor of Peter Carruthers

The State of Physics at the End of the 20th Century

In Honor of Peter Carruthers

Santa Fe, New Mexico, USA

26–29 October 1996

Editors

Fred Cooper
Los Alamos National Laboratory

Ina Sarcevic
University of Arizona

Chung-I Tan
Brown University

Geoffrey West
Los Alamos National Laboratory

World Scientific
Singapore • New Jersey • London • Hong Kong

Published by

World Scientific Publishing Co. Pte. Ltd.

P O Box 128, Farrer Road, Singapore 912805

USA office: Suite 1B, 1060 Main Street, River Edge, NJ 07661

UK office: 57 Shelton Street, Covent Garden, London WC2H 9HE

British Library Cataloguing-in-Publication Data
A catalogue record for this book is available from the British Library.

THE STATE OF PHYSICS AT THE END OF THE 20TH CENTURY: IN HONOR OF PETER CARRUTHERS

ISBN 981-02-3251-9

Printed in Singapore.

CONTENTS

PREFACE

Born in Lafayette, Ind., Peter Ambler Carruthers graduated from the Carnegie Institute of Technology in 1957 and earned a doctorate in theoretical physics in 1960 at Cornell University, where he began his teaching career as one of the youngest faculty members. At Cornell he worked under Dr. Hans A. Bethe, the Nobel laureate. While there, he was among the first physicists to have shown a theoretical inkling of the existence of the hypothetical particles now called quarks.

Carruthers's interests in theoretical physics were unusually broad, from condensed matter to quantum optics and particle physics, from relativistic physics to the theory of strong interactions of elementary particles. He was instrumental in establishing the forefront, fundamental research in the Theoretical Division at Los Alamos National laboratory. As Division Leader, he added 200 scientific personnel and built the Division into the core of the Laboratory and a center of world rank in theoretical physics. More recently, Carruthers focused on the statistics and dynamics of galaxy distributions. He joined the University of Arizona in 1986 as Department Head. He built up the Department, hired 14 new faculty members during his tenure, and brought recognition to the University. For several years he also served as the director of the Center for the Study of Complex Systems at the University. He passed away in August 1997, at the age of 61.

Pete was a renaissance man. He was an accomplished violinist, poet, painter and expert trout fisherman. Pete taught a generation of physicists the art of fly-fishing for trout, long before the movie "River Runs Deep" which popularized this sport for the general public. Pete believed that fly-fishing was more than a sport; it was a way of life. A noted outdoors man once wrote: "Fishermen are searchers. It is true we search for fish, at times with great diligence. But we search also, as men always have, for experiences; ...." We might add, they are like physicists. Pete saw further, and perhaps more deeply, than the rest of us; and for what he taught us, and the humor with which it was conveyed, we can only be grateful.

In October 1996, many of his friends and colleagues gathered in Santa Fe, New Mexico, for a three-day symposium, "Physics at the Beginning of the twenty-first Century", marking his sixtieth anniversary. The first part of this volume collects written versions of some of the talks delivered at the Carruthers Symposium and also includes essays by other friends and colleagues. The second part contains two of Pete's last, unpublished works. The first is based on his experience as a physicist-fly-fisherman in Aspen. The second is a poem by Pete expressing his inner feeling towards the state of theoretical physics during his career.

Fred Cooper
Ina Sarcevic
Chung-I Tan
Geoffrey West

Pete Carruthers Remembered

Sidney Meshkov
Phsics Department, Caltech
Pasedena, California

I have been honored to speak at this conference honoring Peter Carruthers. Pete is both a superb physicist and a fine and trusted friend. I actually met Pete at the Boulder Summer Schools in the middle 60s. We hit it off immediately. When I started coming to Aspen regularly in 1968, Pete, having heard of my interest in salt water fishing, insisted that I learn to fish for trout as a fly fisherman. Consequently, he marched me down to Chuck Fothergill's Fishing store (now the site of Boogies) and outfitted me in full regalia for $234. I caught seven trout that year, at a cost of $33 per trout. My fish amortization got better as the years wore on. A big help in this was my utilization of Pete's superb fly tying ability. He conducted a sideline business of making the most attractive and enduring flies and selling them to his fishing cronies. They were the best! We all vied for the right to buy his Gold Ribbed Hare's Ears, and his Sir Francis Drakes. Would that he could make them for me now.

We hit it off in terms of physics, as well. Pete loved symmetry principles in physics and so do I, so we were kindred souls. In 1965, Harry Lipkin and I invented a symmetries called W-spin and SU(6)W. We were and are to this day quite proud of this work. We found that we were in good company, because Roger Dashen and Murray Gell-Mann, and Pete and K.J. Barnes invented these symmetries quite independently. Pete also wrote a monograph on the uses of SU(3), a subject dear to my heart.

When Pete went to Los Alamos as head of T division, he tried to hire Fred Zachariasen and me to come along with him as his honchos. I didn't do it, probably a great mistake, for Pete built an exciting and vibrant group, many of whom are now honoring Pete. Fred and I both came as visitors and thrived in the exciting atmosphere that Pete built. Great a physicist as Pete was, one could argue that he was at least as good as an organizer. Under his aegis, Los Alamos regained its earlier luster, and became an exciting center of Theoretical Physics.

I saw Pete often, mostly in Aspen, but sometimes when he came to Washington, D.C. to try to extract money from DOE. I was at the National Bureau of Standards, in Gaithersburg, MD, just one exit below Germantown, DOE headquarters, on I-270. Pete would often drop in, as did other buddies of mine, on his forays. He would be wearing his formal money raising clothes, a far cry from his more usual informal garb. Sometimes, he had the time for me to cook Maryland Crab Cakes for him, or to go for a good Chinese dinner, or a Seafood dinner. Pete and I shared a love of some of the finer things in life, including fine food. He was aided and abetted in this pursuit by Lucy. I recall with great delight the many Maurice cooking classes that we attended including Dick and Helene Slansky, Mike and Margaret Simmons, Fred and Nancy Zachariasen, and Steve and Alene Pinsky, Steve and Carla Berry, as well as Pete and Lucy. We would start at 7 PM and finish at about 11 PM. The food and wine were superb, but the companionship and good will were divine.

There are Maurice cooking class tales galore. For one of them, Maurice was going to make truite bleu, requiring live trout. Pete and Fred fished for them and then transported them to the Maurice kitchen in buckets. It worked!

After Pete left Los alamos he built another quality group at the University of Arizona. This was possibly a more difficult task than resurrecting Los Alamos' luster. One of his ventures involved trying to get the SSC to be placed in Arizona. Pete arranged a helicopter ride around the proposed site in Maricopa count, with his usual aplomb.

I left NBS, now called NIST, to work on the SSC, eventually located near Dallas, TX. Despite Pete's efforts on the SSC's behalf, it was killed. For the first time in a long time, I was unemployed. This bothered Pete a lot. Even during Pete's hardest times, he was concerned about me and tried his best to arrange a suitable position for me, a true friend to the end.

Pete Carruthers at Arizona

Daniel Stein
Phsics Department, University of Arizona
Tucson, Arizona

Pete Carruthers arrived at the University of Arizona as the new Head of the Physics Department in the Fall of 1986. During his prior dozen or so years at Los Alamos, he had built the Theory Division into a world class research unit and was later a Senior Fellow. Among his many accomplishments there, he played a role something akin to midwife (or perhaps Godfather) in the modern rebirth of what has come to be called chaos theory. Although trained as a particle physicist under Hans Bethe, Pete had also contributed substantially to other fields, including condensed matter physics. By the time of his move to Arizona, his interests had turned towards the new field of complexity science and he had already played a role in helping to found the fledgling Santa Fe Institute.

This and earlier parts of his story will be told by other speakers at this conference; here, I'd like to focus on Pete's impact on the University of Arizona. He arrived at the University of Arizona with several missions; the first, and most important, was to build the research and teaching strength of the Physics Department. The second was to establish a center for complexity science in Arizona, and the third was to bring the SSC there. On this last item, he managed to bring Arizona into the list of six finalists, no small accomplishment. Although Texas was chosen from that list, in the end, as we all know, there were no winners.

Although no lasting formal center of complexity was established at Arizona, Pete managed to organize and focus the various complexity-related efforts being conducted in disparate departments, and at the same time forged a strong link between the University of Arizona and the Santa Fe Institute, a link which continues to thrive. Pete was a founder of the Complexity Summer School, held each June in Santa Fe, and persuaded me (as a newly hired faculty member, so that my bargaining power at the time was rather limited) to direct the first Summer School. This was, to put it mildly, a daunting task. There were no precedents for such a school. Nobody knew precisely what areas such a school should cover, or who should come. The funding agencies, having an institutional aversion to both interdisciplinary and fundamentally new research, were initially less than enthusiastic. To prove the scientific and administrative feasibility of such an undertaking seemed overwhelmingly difficult that first year. Nonetheless, with strong support from Pete, we got the money, defined the program, invited the participants, and were amazed at how it all came together. This summer we will be holding the tenth annual complexity summer school, which has played a central role both in recruiting young scientists to the field of complexity and in defining new directions for its study.

Pete's primary, and perhaps most lasting, contribution to the University of Arizona was his unprecedented hiring of twelve new faculty in the space of four years. Of course, such rapid and monumental change in any institution initially shocks the system to a state far from equilibrium, and it took awhile for things to relax to a quasi-equilibrium state. But Pete's legacy is that the department is strong and growing, and has taken on a leadership role within the larger College of Science. Such a contribution cannot be measured, but its effects will be long felt.

Pete always had the courage to follow his own vision, and disdained what he saw as the herd mentality of much of the rest of science. He would look to new areas, and their practitioners, while much of the physics community was looking elsewhere. Often, they would eventually come on board, and Pete would then be busy wandering off into a new area where the crowds hadn't yet despoiled the environment. As a Department Head, he believed in leadership rather than mere administration, and wasn't afraid to exercise his authority. While this could cause headaches, it was also the only path to lasting accomplishment. It should be remembered that, at the same time, he was also compassionate, and would help anyone in legitimate need.

Several years ago Pete stepped down as Head (or stepped up, as he would put it), to pursue (without the headaches) his science, music, art, and travel. His many friends around the world all hope that they can give a fraction back to Pete of what he has given us.

State of Physics at the End of the 20th Century: Massive Neutrinos?*

Richard Slansky
Theoretical Division, MS B210
Los Alamos National Laboratory
Los Alamos, NM 87545, USA

A brief review of neutrino masses is presented with focus on how masses might appear in unified models. A fall 1996 status report of the LSND (Liquid Scintillator Neutrino Detector) experiment at Los Alamos is given; the statistical evidence for neutrino oscillations is becoming stronger. A summary of a unified model based on $SO(18)$ shows possible complications in understanding neutrino masses.

1 Introduction and General Issues

It is a pleasure to be part of Pete Carruther's 61st birthday celebration. Pete was the Theoretical Division Leader when Murray Gell-Mann, Pierre Ramond and I looked at the "problem" of neutrino masses in the late 70's.[1] We tried to understand neutrino masses in unified models, especially the $\mathcal{G} = SO(10)$ model where each "family" of left-handed fermions is assigned the spinor **16**.[2] ($\mathcal{G}$ is the gauge group of the theory; representations are in boldface type.) Here it is natural for neutrinos to have Dirac masses of nearly the same size as other masses in the family. In this setting it is hard to understand masslessness of neutrinos; comparatively small masses are easier.

Understanding the empirical patterns of fermion masses, the angles between the mass and current eigenstates of the quarks, and the analogous mixing angles for the leptons for nonzero neutrino masses all continue to be challenging. The theory of the strong and electroweak interactions based on $\mathcal{G} = SU(3)^C \times SU(2)^W \times U(1)^W$ provides little constraint on these parameters. Quantum loops do not renormalize zero neutrino masses because of the chiral symmetry of zero-mass fermions, and the mixing angles remain zero. However, there is growing evidence that neutrino masses are small but nonzero.[3]

The span of the fermion masses is huge. The ratio of the top-quark mass to the electron mass is about 3.5×10^5. Evidence for neutrino oscillations suggests nonzero neutrino masses,[4] so there may be larger ratios. Radiative corrections may be responsible for some large mass ratios, but the overall pattern does not appear simple. Masses do not seem to be directly proportional to vector-coupling coefficients of $\mathcal{G}$, even after renormalization to low energies.

This talk begins with a brief summary of general issues concerning neutrino masses, followed by a status report of the LSND measurement of neutrino oscillations.[4] Finally, we comment on some complications in the fermion mass matrix in a unified model based on $\mathcal{G} = SO(18)$, where all the known left-handed fermions are assigned to one irreducible representation.

In the $\mathcal{G} = SO(10)$ model the 45 known left-handed fermionic degrees of freedom are assigned to three **16**s:

$$\mathbf{f_L} = \mathbf{16_1} + \mathbf{16_2} + \mathbf{16_3} \ , \tag{1}$$

with one **16** for each family of quarks and leptons and $\mathbf{f_L}$ is the gauge group representation of the left-handed fermions. For example, $\mathbf{16_1}$ in Eq. (1) contains the left-handed electron, electron neutrino, positron, plus the three colors each of up and down, quarks and antiquarks. Each of these 15 degrees of freedom has a $\Delta I^W = \frac{1}{2}$ mass except for the neutrino, which is assumed to have zero mass in the usual statement of the standard model. By itself, this one neutral degree of freedom could have a $\Delta I^W = 1$ Majorana mass that violates lepton number by 2 units. The 16-th degree of freedom of the **16** is neutral under the electroweak interactions.

In $SO(10)$ the neutrino mass matrix is that of a four-component fermion with lepton-number violating terms allowed. This is conveniently analyzed in the Weyl basis with Majorana spinors. A Majorana spinor is a Dirac spinor satisfying $\psi = \psi^c$, or $\overline{\psi}_R = \psi_R$ and $\overline{\psi}_L = -\psi_L$ in the Weyl basis, so ψ has just two physical degrees of freedom. In terms of a real field ρ the Majorana fields are

$$\psi^T = \left(\rho_R, \, -\rho_L^\dagger, \, \rho_L, \, \rho_R^\dagger\right), \quad \overline{\psi} = -\left(\rho_L^\dagger, \, \rho_R, \, \rho_R^\dagger, \, -\rho_L\right) . \tag{2}$$

Each neutral fermion of the **16** contributes a component to ρ_L, so $\rho_L^T = (\nu_L, \overline{\nu}_L)$, each with different electroweak quantum numbers.

A Majorana spinor does not necessarily describe a particle that is its own antiparticle. The charge-conjugation operator $\mathcal{C}$ is defined to reflect one component of ρ_L to another. Thus, if a spinor e_L^- describes a left-handed electron (and the right handed positron), then a second Majorana spinor e_L^+ can be introduced that describes a left-handed positron. These spinors can be put together to form a Dirac electron using $\mathcal{C}$. If e_L^- and e_L^+ are in the same irreducible representation of $\mathcal{G}$, $\mathcal{C}$ is associated with $\mathcal{G}$.[5] Similarly, it is convenient to describe the two neutral fermions of the **16** as two Majorana spinors.

The most general mass term with Majorana spinors is

$$\mathcal{M} = \frac{1}{2} : \overline{\psi}(S + iP\gamma_5)\psi : = \left(\rho_R^\dagger, \, \rho_L^\dagger\right) \begin{pmatrix} 0 & M \\ M^\dagger & 0 \end{pmatrix} \begin{pmatrix} \rho_R \\ \rho_L \end{pmatrix} , \tag{3}$$

where the explicitly hermitian form follows from Eq. (2). The colons around the mass operator signify normal ordering, including a sign change when exchanging a pair of fermion operators. M is the symmetric matrix,

$$M = M^T = -\frac{1}{2}(S - iP) - \frac{1}{2}(S - iP)^T. \tag{4}$$

The diagonalization of the mass matrix would appear to require an analysis of the matrix in Eq. (3). However, by a theorem of Shür it is only necessary to compute the eigenvalues of the matrix $M^\star M$: the matrix $M^\star M$ is Hermitian and positive definite for symmetric M. The eigenvalues of the full mass matrix are plus/minus the square roots of the eigenvalues of $M^\star M$. In the case that CP is conserved, M is also a real matrix, and so the eigenvalues of $M^\star M$ are simply the squares of the eigenvalues of M. Thus, a study of the mass operator is a study of the symmetric matrix M.

The mass matrix of a four-component neutrino for the $\rho_R^\dagger M \rho_L$ term of Eqs. (3) and (4) is

$$\rho_R^\dagger M \rho_L = \left(\nu_R^\dagger, \overline{\nu}_R^\dagger\right) \begin{pmatrix} 0 & m \\ m & M_0 \end{pmatrix} \begin{pmatrix} \nu_L \\ \overline{\nu}_L \end{pmatrix}. \tag{5}$$

The parameter "0" is the $\Delta I^W = 1$ mass of the left-handed neutrino (it is not required to be zero, but is certainly expected to be quite small), m is the Dirac mass for the neutrino, and M_0 is the $\Delta I^W = 0$ right-handed neutrino Majorana mass that could be huge, perhaps 10^{10} GeV. The two neutrino like degrees of freedom are mixed by the Dirac $\Delta I^W = \frac{1}{2}$ mass, m. If the $\Delta I^W = 1$ mass matrix element is zero, then the observed neutrino mass can be very small. Solving the mass eigenvalue equation yields approximately m^2/M_0 for the smaller-mass solution. The neutrino has a mass smaller than typical Dirac masses in the **16** by the ratio m/M_0.

Suppose the left-handed fermions are assigned to the representation $\mathbf{f_L}$ as in Eq. (1). By Eq. (4) the mass matrix couples to $(\mathbf{f_L} \times \mathbf{f_L})_S$, where the subscript S indicates the symmetric part of the product. Both the symmetric and antisymmetric parts of the product

$$\mathbf{16} \times \mathbf{16} = \mathbf{126}_S + \mathbf{10}_S + \mathbf{120}_A \tag{6}$$

can contribute several times each to $(\mathbf{f_L} \times \mathbf{f_L})_S$ in a three-family model, depending on the complexity of the mass operator. $SO(10)$ provides little guidance on the values of the mass-matrix elements.

There have been several approaches to finding a more constrained mass matrix. There may be a family symmetry that ties together the three families and further constrains the mass matrix. If such a symmetry is continuous, one might ask how it is related to some unification of the standard model. A simple possibility for providing such constraints is for $\mathbf{f_L}$ to be a single, irreducible representation of $\mathcal{G}$.

Several unification schemes have been suggested where $\mathbf{f_L}$ is irreducible.[1] The **1728** of $\mathcal{G} = E_6$ could contain the known 45 degrees of freedom. It contains many unobserved states with large values of weak isospin, color and electric charge. There is no obvious reason for all of these masses to be larger than the known quark masses. Another example is the 256-dimensional spinor representation of $\mathcal{G} = SO(18)$. In this case there are still over 200 states that have not been observed. At least their weak and strong quantum numbers do not have large weak isospin, color or electric charge as in the **1728**.

The additional degrees of freedom must be sufficiently hidden. The decomposition of the **256** into $SO(10)$ representations yields eight **16**s and eight $\overline{\mathbf{16}}$s. We suppose there is a "color-prime" force that hides all the $\overline{\mathbf{16}}$s and five of the **16**s. The experimentally observed three families are then color-prime singlets. The properties of the mass matrix under $\mathcal{C}$ will be described in Sec. 3.

2 Neutrino Oscillations and LSND

The experimental effort to look for small masses by searching for oscillations depends on masses and interaction currents having different origins so they can be out of alignment. Interactions are defined by the gauge symmetries and the coupling of vector bosons to the symmetry currents. The origin of mass is in the symmetry breaking mechanism. The CKM matrix measures this misalignment for the quarks.

Also in the case of neutrinos the family identified by the interaction producing the neutrino does not have to line up with the neutrino mass eigenstates. From Eq. (3) the CKM matrix for three neutrinos and three antineutrinos is six-by-six. A neutrino produced in a weak interaction may change its family due to oscillation, for example, a $\overline{\nu}_\mu$ from μ^+ decay, may oscillate to a $\overline{\nu}_e$ if the masses of the $\overline{\nu}_\mu$ and $\overline{\nu}_e$ differ and the mass matrix does not line up with the weak interactions. The crucial parameters describing the oscillation length is $\Delta m^2 = m^2_{\nu\mu} - m^2_{\nu e}$ and the amplitude of the oscillation is $\sin\theta$ where θ measures the misalignment of the weak interactions and the mass matrix.

At the time of this celebration of Carruther's birthday, the LSND experiment at Los Alamos was half way through its second run. The data here includes the first run (1992-1995) and data from the second run. Within statistics the data from the two runs are consistent with one another, and the overall statistics of the experiment are improved by combining the data. In the second run the tank was placed slightly closer to the beam stop, the length of the decay channel was changed, and other systematics of the experiment changed slightly.

Most of the neutrinos produced by the secondary interactions and decays resulting from the collision of 800 MeV protons of the LAMPF beam with the beam stop are due to π^+ and μ^+ decay. Only 11% of the pions produced are π^-, and 95% of these are absorbed by the beam stop before they can decay into μs and νs. Thus, the production by usual weak processes of $\overline{\nu}_e$s is greatly suppressed. The LSND collaboration finds that [4]

$$\overline{\nu}_e/\overline{\nu}_\mu = 7.5 \times 10^{-4} . \tag{7}$$

It is not sufficient simply to look for an excess of $\overline{\nu}_e$s; there must be a positive signal of $\overline{\nu}_e$ interactions with the right energy distribution. Particle identification and energy are measured by using both the scintillation and Cerenkov signals in the LSND tank. (The tank is full of baby oil with some scintillator added to it and observed by an array of phototubes.)

The neutrino-oscillation-event signature described here is from the reactions where the $\overline{\nu}_\mu$ from μ^+ decay has oscillated to $\overline{\nu}_e$, and the $\overline{\nu}_e$ is detected by the reaction,

$$\overline{\nu}_e + p \longrightarrow e^+ + n, \tag{8}$$

where the positron is directly observed. There is a positron background at lower energies, but for positrons above 36 MeV, there is a significant signal. Before other cuts on the data, there is an excess of 63.3 ± 20.1 events.

The additional signal is the identification of the neutron in Eq. (8) by observing a γ with a time delay of 186 μsec from the reaction

$$n + p \longrightarrow d + \gamma(2.2\ \text{MeV}). \tag{9}$$

With a fairly tight restriction on the γ signal, the excess of $\overline{\nu}_e$ events becomes [4]

$$\text{Excess} = 17.4 \pm 4.7\ \text{events}. \tag{10}$$

The probability of this being a statistical fluctuation is 4.2×10^{-8}.

It is customary in neutrino oscillation experiments to show solutions (or excluded regions) on a plot of Δm^2 versus $\sin^2 2\theta$. This oscillation analysis assumes that only two neutrino types participate in the oscillation, but the analysis gives some indication where the parameters are likely to lie after additional data is acquired and all three neutrino types are included.

The strongest experimental result is the result given in Eq. (10), which identifies that there is an anomalous source of $\overline{\nu}_e$s. The solution region for the parameters Δm^2 and $\sin^2 2\theta$ in LSND includes a positive signal that lies outside the exclusion regions of Brookhaven's E776 experiment, the Karmen data and the reactor experiment at Bugey. (All three of these experiments place restrictions on the Δm^2 and $\sin^2 2\theta$ parameters, with no evidence of a positive signal.) LSND has a positive solution to its signal that lies on a line between $\Delta m^2 \approx 2\ eV^2$ and $\sin^2 2\theta \approx 0.002$ and $\Delta m^2 \approx 0.3\ eV^2$ and $\sin^2 2\theta \approx 0.03$. Other solutions lie in excluded regions by other experiments.

3 The Spinor of $SO(18)$

The Lie algebra of $SO(18)$ is rank 9 and has 153 basis elements. In the Weyl basis, there are 9 diagonal operators, 72 lowering operators (which lower the quantum numbers of a state) and the corresponding 72 raising operators. Although this is a large algebra, the analogous results for $SO(10)$ are well-known and add intuition about the rather large representations of $SO(18)$.

The basis vectors of the spinor representations of $SO(6+4n_f)$ all have unit multiplicity (n_f is number of families); there is just one Hilbert-space vector for each set of $SO(6+4n_f)$ quantum numbers, so that eigenvalues of the diagonal operators completely label the basis states. [5]

The three-family model is based on the spinor **256** of $SO(18)$. [1] The $SO(8)\times SO(10)$ branching rule of the **256** is $(\mathbf{8_s},\mathbf{16})+(\mathbf{8_c},\overline{\mathbf{16}})$. The $\mathbf{8_s}$ has the $SO^f(3)\times SO^{c'}(5)$ decomposition, $(\mathbf{3},\mathbf{1})+(\mathbf{1},\mathbf{5})$; the conjugate spinor $\mathbf{8_c}$ has the decomposition $(\mathbf{2},\mathbf{4})$. Thus, the $SO^f(3)\times SO^{c'}(5)\times SO(10)$ decomposition of the **256** is

$$\mathbf{256} = (\mathbf{3},\mathbf{1},\mathbf{16}) + (\mathbf{1},\mathbf{5},\mathbf{16}) + (\mathbf{2},\mathbf{4},\overline{\mathbf{16}}). \tag{11}$$

The $SO^f(3)$ is assumed to be the "family symmetry" and the $SO^{c'}(5)$ is a confining "color-prime" symmetry. Without further discussion we simply assume that the $SO^{c'}(5)$ confines so all but the 48 states in $(\mathbf{3},\mathbf{1},\mathbf{16})$ are manifested as bound states of color-prime. [1] We also assume that only the color-prime singlets have been observed experimentally. If there is any reality to this model, the composites of non-singlet color-prime states have masses above present experimental observation, even though many masses have $\Delta I^W = \frac{1}{2}$.

The next piece of $SO(18)$ group theory concerns the representations that couple to the mass matrix, $(\mathbf{256}\times\mathbf{256})_S$, where the subscript S indicates that the mass matrix couples only to the representations in the symmetric part of the tensor product, as derived in Eq. (4),

$$(\mathbf{256}\times\mathbf{256})_S = \mathbf{24310} + \mathbf{8568} + \mathbf{18}. \tag{12}$$

With only three representations there would appear to be strong constraints on the mass matrix. Nevertheless, there are many components in each of these representations that might have nonzero vacuum values after radiative corrections are included.

The part of the analysis discussed here is whether there exists for $SO(18)$ a charge-conjugation operator $\mathcal{C}$ that gives a simple classification of the mass-matrix elements. For example, the Dirac masses are all even under $\mathcal{C}$. Although $SO(18)$ has the possibility of defining a number of charge conjugations, none of these is particularly attractive.

To define the charge conjugation, divide the Lie algebra into two parts: the symmetric subalgebra $\mathcal{S}$, and the remaining operators $\mathcal{A}$. A symmetric subalgebra $\mathcal{S}$ is defined by the set of commutation relations,

$$[\mathcal{S},\ \mathcal{S}] \subseteq \mathcal{S},\ [\mathcal{S},\ \mathcal{A}] \subseteq \mathcal{A},\ [\mathcal{A},\ \mathcal{A}] \subseteq \mathcal{S}. \tag{13}$$

Thus, we can define an automorphism on the full Lie algebra as

$$\mathcal{C}(\mathcal{S}) = \mathcal{S},\ \ \mathcal{C}(\mathcal{A}) = -\mathcal{A}. \tag{14}$$

It is obvious that the sign changes of Eq. (14) leave the commutation relations in Eq. (13) unchanged. It is straight forward to derive all the symmetric subalgebras, the action of $\mathcal{C}$ of the vectors of a representation, and to count the number of negative and positive eigenvalues of $\mathcal{C}$.[5]

$SO(18)$ has four candidate charge conjugations that flip the signs of the eigenvalues of four or more diagonal operators. These are:

- $\mathcal{S} = SO(8) \times SO(10)$, where $\mathcal{C}$ flips the signs of eight operators. In this case $\mathcal{C}$ looks exactly like the $\mathcal{C}$ for $SO(10)$. In addition it flips all the signs in the diagonal operators in $SO(8)$. ($\mathcal{C}$ of $SO(10)$ flips the signs of four of the five diagonal operators, including electric charge and the two diagonal color operators. It cannot flip all five diagonal operators and reflect the **16** to itself.)

- $\mathcal{S} = SO(6) \times SO(12)$, where $\mathcal{C}$ flips the signs of six operators. This $\mathcal{C}$ also acts exactly like the $\mathcal{C}$ of $SO(10)$. It flips just 2 of the signs in the $SO(8)$.

- $\mathcal{S} = SO(4) \times SO(14)$, where $\mathcal{C}$ flips the signs of four diagonal operators. It acts exactly as the $\mathcal{C}$ of $SO(10)$. It flips no sign of the $SO(8)$.

- The most interesting candidate is associated with the symmetric subgroup $\mathcal{S} = SU(9) \times U(1)$, which flips four operators. Moreover, this is the only case that does not pair up half of states in the **256** with the other half. However, it proves impossible to orient the subgroup in $SO(18)$ so that it flips the diagonal color operators and electric charge. The proof of this claim is based on the fact that $\mathcal{S} = SU(5) \times U(1)$ is a symmetric subgroup of $SO(10)$, where $\mathcal{C}$ flips the sign of only two of the diagonal operators. The final result can then be built up to the $\mathcal{S} = SU(9) \times U(1)$ case for $SO(18)$. Thus, this case is not a candidate.

In the first three cases no state is transformed outside its **16**. The Majorana masses for the neutrinos are transformed into one another, so that $\mathcal{C}$ is maximally violated by the mass matrix as can be seen in Eq. (5).

This analysis leaves us with the conclusion that the mass matrix in $SO(18)$ has mixed behavior under $\mathcal{C}$. In every case the $\mathcal{C}$ exchanges ν_L with $\overline{\nu}_L$ within each $SO(10)$ family. Thus, the behavior of the mass matrix is not simplified by the charge-conjugation analysis. This does not guarantee the failure of the $SO(18)$ model, but does not help clarify the origin of the large Majorana mass in Eq. (5).

* Work performed in part under the auspices of the U. S. Department of Energy, under Contract No. W-7405-ENG-48.

References

1. M. Gell-Mann, P. Ramond and R. Slansky in *Supergravity*, ed. P. van Nieuwenhuizen and D. Z. Freedman (North Holland, Amsterdam, 1979), 315.
2. H. Fritzsch and P. Minkowski, Ann. of Phys. (N.Y.) **93** 193, (1975); H. Georgi in *Particles and Fields (1974)*, ed C. E. Carlson (AIP, New York, 1975).
3. R. S. Raghaven, *Science* **267**, 45 (1995).
4. C. Athanassopoulos *et.al.*, *Phys. Rev. Lett.* **75**, 2650 (1995); C. Athanassopoulos *et.al.*, *Phys. Rev.* C**51**, R1065 (1995). I am grateful to Bill Louis for providing me with data from the second run of LSND.
5. R. Slansky, *Phys. Reports* **79**, 1 (1981).

$\lambda\phi^4$ Field Theory: Still a Few Kinks in the System

David Campbell
Department of Physics, University of Illinois
Urbana, Illinois

Mentioning the word "web" today among a group of physicists immediately evokes images of computers, cybermillonaires, and frustrating traffic jams on the information "super highway." But there is another, older, more important "web" in physics and science, as in life: the complex array of relationships among scientists themselves, their research, their lives, the institutions they create and populate, and the progress of science. Ultimately, this web links all scientists and all of science, often connecting seemingly disparate and distant nodes via surprisingly short paths. Picking at random any given strand, one can explore how the various filaments of this human/scientific web intertwine, how nodes condense, and how links are created or broken. The results of this exploration can be fascinating and rewarding, offering rich insights not only into science but also into "Life, the Universe, and Everything" [1].

In this contribution to the volume honoring the contributions and memory of Peter Carruthers, I have chosen one particular small strand—my own studies of the one-dimensional (1D) $\lambda\phi^4$ field theory—and will trace its path through this "human" web, exploring (briefly) some of its many links to other research results and (more extensively) the connections to Pete, Los Alamos, T-Division, and the Center for Nonlinear Studies (CNLS). As we shall see, this strand extends for the whole quarter century that I knew Pete, and its links to other filaments form a highly nonlinear network, and often kinked (or even kinky). And, as the title suggests, "there are still a few kinks in the system." [1]

As I recall, Pete and I met briefly during my first visit to Aspen, in 1973, and the context (appropriately enough) was fishing. I was a rank amateur fly fisherman, with just enough sense to ask who the experts were, and was told that there was one master: Pete. So I sought his advice and got that—Hunter Creek for beginners—and, more importantly, received a small cache of his flies. Elsewhere in this volume Sid Meshkov has sung the praises of Pete's Gold-Ribbed Hare's Ears and Sir Francis Drakes, but for me the old

[1] All puns and *double entendres* should be assumed intentional, even the ones I don't get.

reliable was Pete's Rio Grande King. One fine Aspen day, on a beaver pond somewhere off the road to Independence Pass [2], I caught four cutthroats, each over 12", on four successive casts of one of Pete's Kings. In my tackle box, I still keep a couple them—Pete's Kings, that is, not the cutthroats—much trout-bitten and unusable, just for their magic. At the end of that summer, we drove south to Santa Fe, and as we passed Los Alamos I pointed out the Lab. My wife uttered some variant of "it's a beautiful place, but I wouldn't want to live here": a quarter of a century later, she won't consider living anywhere else.

The circumstances of my coming to Los Alamos, as the first J. R. Oppenheimer Fellow in August 1974, reveal several aspects of Pete's leadership and managerial skills and offer instructive lessons to would-be administrators. First lesson: always have a plan B. When Pete went before the Oppenheimer selection committee to argue my case, he actually had *two* candidates— my friend Bill Colglazier, who was also at the Institute for Advanced Study at that time, and me—and he waited to see which way the committee was leaning before pulling out one of his two pre-prepared "X is the best thing since sliced bread" speeches. When he sensed that the committee was lean toward my candidacy, he set "X = Campbell" and proceeded. Second lesson: be ready for the grand take-out. Both Bill and I were high-energy particle theorists, and then (as now) some members of Lab wondered about the relevance of such research at Los Alamos. When a selection committee member voiced concern on this point, Pete straightened his tie [3] looked the individual directly in the eye, and said: "If Robert Oppenheimer were alive today, he would call himself an elementary particle physicist." Game, set, and match to Carruthers [4], and Oppenheimer Fellowship to Campbell.

I arrived in Los Alamos in mid-August 1974, one of the "new crew" of particle physics types (Dick Slansky, Mitchell Feigenbaum, and Geoff West) recruited by Pete in his first wave of hiring. I came alone, in advance of my family, and discovered that, with Pete's attention focused on more important matters, no reservation had been made for me in town. The only place available for me until I could move into my rented sabbatical house in early September was at the bottom of Frijoles Canyon, in the cabins built in Bandelier National Monument by the WPA. Twenty miles commute to work

[2] You don't expect me to tell you where, do you?

[3] As pointed out by Sid Meshkov in his contribution to this volume, Pete was capable of dressing for the occasion, and this occasion, he told me later, required a tie.

[4] That brings up another aspect of Pete's persona: his approach to tennis. I think I'll leave that to one of his longer-time tennis partners, Mike Simmons or Fred Zachariasen.

every day ! No telephone ! I was livid. I remained livid for about 12 hours, until I woke up to the glorious canyon sunrise and fell in love with New Mexico. So Pete had inadvertently provided a third lesson: don't sweat the small stuff—things have a way of working out, if you focus on the important matters.

Just prior to my arrival in Los Alamos, motivated by the celebrated papers of Dashen, Hasslacher, and Neveu (DHN) [2]), I had begun working on the semiclassical quantization of a 1D field theory consisting of fermions coupled to a nonlinear boson field, the Lagrangian for which was given by

$$\mathcal{L} = \int dx \left[\frac{1}{2}\left(\frac{\partial \phi}{\partial t}\right)^2 - \frac{1}{2}\left(\frac{\partial \phi}{\partial x}\right)^2 - \frac{1}{4}(\phi^2 - 1)^2 \right]. \tag{1}$$

The semiclassical analysis required knowledge of the solutions to the corresponding classical field equation, which reads

$$\frac{\partial^2 \phi}{\partial t^2} - \frac{\partial^2 \phi}{\partial x^2} - \phi + \phi^3 = 0. \tag{2}$$

Apart from the trivial (degenerate vacuum/ground state) solutions $\phi_\pm = \pm 1$, this equation admits the now famous "kink" soliton solution, which because of its centrality to our discussion I write explicitly:

$$\phi_K = \tanh \frac{\gamma(\mathrm{x} - \mathrm{vt})}{\sqrt{2}} \tag{3}$$

with $\gamma = 1/\sqrt{1 - v^2}$. There is also an antikink, with $\phi_{\bar{K}} = -\phi_K$. Each of these solutions represents a spatially localized distribution of energy and momentum and in the quantum theory corresponds to a novel, "nonlinear" eigenstate, not accessible via perturbation theory from the vacuum.

Eventually, I was able to find new semiclassical solutions to the coupled fermion-boson problem [3], to show how they could be related to (models of) particle and nuclear physics (the kink, with its zero energy fermion state occupied, is just the 1D version of the SLAC bag) [4]), and somewhat later to apply similar solutions to explain the properties of solitons and polarons in conducting polymers [5, 6]. But to pursue these filaments of the web would take us too far from the main strand. Interested can follow up through the references.

At the same time I was focusing on nonlinear "soliton" solutions to the semiclassical equations of ϕ^4 and related field theories, Mitchell Feigenbaum had begun his classic work on the period-doubling transition to chaos. We talked about our respective progress on a regular basis, unwittingly bringing together what were to become two "paradigms" defining the initial research agenda of the Center for Nonlinear Studies (CNLS). Before discussing the CNLS and Pete's role in its formation, let me follow the ϕ^4 strand to the next node, especially since this leads to another "classic" Los Alamos tale.

The semiclassical studies involving ϕ^4 solitons led me naturally to discussions with Chuck Wingate, who (moonlighting from his daytime assignments in (then) TD-Division) was interested in numerical simulations of interactions of solitons in *non*-integrable field theories. Soon Chuck and I had some interesting preliminary results showing a very unexpected resonance structure in kink-antikink interactions. When I was invited to give an external talk, I suggested that topic. In those days (as now), one needed to get formal clearance from the Lab's classification office for all talk abstracts. So I headed over to old T-Division office in the Administration Building to have Pete's secretary (Lois MacFarland) type (*sic*) up the forms. It happened that Mitchell was in the office at the same time. I dictated the title to Lois, slowly pronouncing each word in the phrase "Kink-Antikink interactions in $\lambda\phi^4$ Field Theory..." and without a hint of hesitation Mitchell continued "....and Applications to Nuclear Weapons." Lois just kept typing—we had to start with a new form.

Several other contributors have commented on Pete's excellent taste in physics and his uncanny ability to sense and facilitate new developments. The critical role Pete (together with Mike Simmons) played in the formation of the CNLS illustrates this perfectly. While several of us junior staff members (Mitchell, myself, Alan Bishop, Darryl Holm, Mac Hyman, and Basil Nichols) toiled at the grass roots level, Pete and Mike worked the senior administration, arranging for crucial input at the Director's level from the "Polish masters" Stan Ulam and Mark Kac and providing small amounts of initial seed funding and large amounts of encouragement. Characteristically, once he had enabled the creation of the CNLS (in October 1980), Pete let it develop on its own, maintaining an active interest in the science but avoiding any micro-management. Lesson number four for future czars: facilitate and enable, don't micro-manage.

Returning to the ϕ^4 strand, after considerable effort a compelling heuristic explanation of the resonances in kink-antikink scattering emerged. Our results showed that the resonances arose via a "resonant energy exchange mechanism" between collective coordinates representing the kink's center of mass motion and shape oscillation modes localized around the kink [7]. We were able to show that this mechanism applied to many other *non*-integrable classical field theories, and that in the absence of localized shape modes, the resonances did not occur [8, 9, 10]. More recently, others have considerably extended and formalized these collective coordinate approaches, applying them to a wide variety of nonlinear excitations, including Skyrmions, monopoles, and vortices [11, 12] (under fancy monicker "moduli spaces").

A separated and complex scientific strand tied to our kink-antikink collision studies was the search for an exact ϕ^4 "breather": *i.e.*, an exponentially localized (in space), periodic (in time) exact solution to Eqn. (2) above. For the sine-Gordon equation

$$\frac{\partial^2\theta}{\partial t^2} - \frac{\partial^2\theta}{\partial x^2} + \sin\theta = 0, \tag{4}$$

the explicit analytic form of the breather solution is known to be

$$\theta_B(x,t) = 4\tan^{-1}\left(\frac{\varepsilon\sin(t/\sqrt{1+\varepsilon^2})}{\cosh(\varepsilon x/\sqrt{1+\varepsilon^2})}\right), \tag{5}$$

for any value of ε. Although no one expected a closed form analytic expression for the ϕ^4 breather, the questions remained, "Did such an exact solution exist, and could it be found numerically?" The search for the ϕ^4 breather was strongly motivated by two observations: (1) the presence of long-lived, localized, large amplitude oscillatory states formed for certain velocities in kink-antikink collisions [7]; and (2) the existence, within perturbation theory about the ground state, of small amplitude "breathers." This second observation was first made by DHN in their classic article [2], where they presented an analytic expression for the ϕ^4 breather to lowest orders in an expansion in ε (the deviation from the ground state solution) and argued that one could show *to all orders in* ε that this breather was exponentially localized and time-periodic. Specifically, using DHN's arguments and analysis, one can calculate the form of the putative ϕ^4 breather to third order in ε:

$$\phi_B^{(3)} = 1+\frac{2\varepsilon}{\sqrt{3}}\text{sech}\xi\cos\tau-\varepsilon^2\text{sech}^2\xi+\frac{\varepsilon^2}{3}\text{sech}^2\xi\cos 2\tau+\frac{\varepsilon^3}{6\sqrt{3}}\text{sech}^3\xi\cos 3\tau, \tag{6}$$

where $\xi = \frac{\varepsilon\sqrt{2}x}{\sqrt{1+\varepsilon^2}}$ and $\tau = \frac{\sqrt{2}t}{\sqrt{1+\varepsilon^2}}$. There were two competing schools of thought regarding the existence of a true ϕ^4 breather. The computational physicists, who had studied numerical solutions to partial differential equations (PDEs) and had seen seemingly stable, localized, oscillatory "blobs" in theories like time-dependent Hartree-Fock, were convinced that one could "tune" parameters such that an exact breather solution could be found numerically. But the experts on integrable systems theory argued that the existence of the higher-order conservation laws in integrable systems was crucial to the existence of higher-order solitons (the sine-Gordon breather can be thought of as a two-soliton bound state), so that if a theory was *not* integrable (and ϕ^4 is not), it should not have any breathers.

For many years, several collaborators (including Chuck Wingate, John Negele, and Michel Peyrard) and I struggled to find conclusive numerical evidence either for or against the ϕ^4 breather conjecture. Our numerical results, particularly for larger amplitudes, strongly suggested the existence of of breathers in (spatially) *discretized* approximations to ϕ^4, where (in the simplest case) we approximated the continuum spatial partial derivative by

$$\frac{\partial^2 \phi}{\partial x^2} \sim \frac{1}{(\Delta x)^2} \left(\phi(x + \Delta x) - 2\phi(x) + \phi(x - \Delta x)\right). \tag{7}$$

For finite Δx, this reduces the PDE to a set of N coupled nonlinear oscillators, where $N\Delta x = L$, the length of the spatial interval studied. Recovering the continuum limit requires taking $\Delta x \to 0$ and $N \to \infty$ such that L is constant. When we attempted to take this limit, we found many indications of substantial subtleties, which we discussed with a number of experts. Finally, we convinced Harvey Segur to look at the problem, and together with Martin Kruskal [13], he was able to show that there were terms "beyond all orders" in ε, so that exact solution was

$$\phi_{True} = \phi_b + \exp\left(-1/\varepsilon\right) \cos(k_c x - \omega_c t) + \cdots, \tag{8}$$

where $\omega_c = 2\sqrt{2}(1 - \varepsilon^2/2)$, $k_c = \sqrt{\omega_c^2 - 2}$, and c_1 and c_2 are two (non-zero) constants. Hence ϕ_{True} is *not* localized in space and in the continuum ϕ^4 theory there is in fact no true "breather" solution [1] This result inspired attempts to prove rigorously the absence of breathers in *non*-integrable continuum models; after several partial results, including a couple of surprising missteps, this result was finally proved [14].

Michel and I summarized our results in a long article published in an AIP conference proceedings (and hence generally overlooked) [15]. We found a complex "fractal"/ Arnold tongue-like structure of stability regions as a

[1]Such (infinite energy) solutions, with very small amplitude phonon "wings" going out to infinity, have been given the name "nanopterons," in my view one of the least successful of the many neologisms associated with solitons.

function of discreteness and breather frequency. Clearly, there were regions in which localized breathers on discrete lattices appeared to be stable, and we argued that this stability might persist even on infinite lattices, when the phonon spectrum (still of finite width because of the discreteness) becomes dense. Thus we had clear hints of the existence of time-period, "intrinsically localized modes" in discrete—but not continuum—theories. For various reasons— in my case, primarily related to my move to Urbana and the increased administrative responsibilities—we left this problem at that stage, just before the vital insight of Aubry and Mackay [16] who, starting from the anticontinuum limit ($\Delta x \to \infty$, in our notation) of an isolated single-site discrete breather, were able to prove the existence of stable discrete breathers in infinite lattices for sufficiently weak coupling to adjacent lattice sites (*i.e.*, sufficiently large Δx) and to use a constructive method of proof to continue the solutions numerically to finite coupling. This important development also provided firm analytic footing for earlier studies of intrinsically localized modes (ILM) [17] and led to an explosion of interest in these discrete nonlinear objects [18].

One final strand in our scientific web is also tied to the *discretized* version of ϕ^4 theory: namely, the existence of a quantum critical point (phase transition at zero temperature) as a function of "discreteness" (at fixed other parameters). We were able to show, using analytic estimates based on quantum corrections to the energies of ϕ^4 kinks which we then confirmed by quantum Monte Carlo simulations, that the broken symmetry of ϕ^4 theory could be restored by sufficient "discreteness" and to apply this result to the behavior of hydrogen-bonded systems under pressure [20].

To ensure that our tour of the "human" web comes full circle, I will end with another Pete fishing story. The last time Pete and I fished together was in October 1990 (or 1991) in Aspen, and my younger son (Micha) was along. Pete and Micha took off in one direction and I in another, and when we gathered for lunch, Micha had by far the largest fish, a fine 14-incher. He had caught it in a riffle that Pete had pointed out to him. I think often of this as an excellent metaphor for Pete's approach to many things in his life: he was able to facilitate and encourage the success of others, and to take real pleasure in these successes. He made certain that his shadow did not obscure his junior colleagues and that the institutions he set up could succeed when he moved on. He thus left a lasting legacy. This is the fifth, final, and most important lesson.

References

[1] Douglas Adams, "Life, the Universe, and Everything," Volume 3 in "The Hitchhiker's Guide to the Galaxy," (Ballantine Books, Reprint Edition, 1995).

[2] R. F. Dashen, B. Hasslacher, and A. Neveu, "Particle Spectrum in Model Field Theories from Semiclassical Functional Integral Techniques," *Phys. Rev. D* **11**, 3424-3459 (1975).

[3] D. K. Campbell, "Exact Classical Solutions of the Two Dimensional Sigma Model," *Phys. Lett. B* **64**, 187-190 (1976); D. K. Campbell and Y. T. Liao, "A Semi-Classical Analysis of Bound States in the Two-Dimensional Sigma Model," *Phys. Rev. D* **14**, 2093-2116 (1976).

[4] D. K. Campbell, "Nuclear Physics in One Dimension," pp. 673-716 in *Nuclear Physics with Heavy Ions and Mesons*, R. Balian, M. Rho, and G. Ripka, eds., (North Holland, 1978).

[5] D. K. Campbell and A. R. Bishop, "Solitons in Polyacetylene and Relativistic Field Theory Models," *Phys. Rev. B* **24**, 4859-4862 (1981).

[6] D. K. Campbell, A. R. Bishop, and M. J. Rice, "The Field Theory Perspective on Conducting Polymers," pp. 937-966 *Handbook of Electrically Conducting Polymers*, T. A. Skotheim, ed., (Marcel Dekker, 1986).

[7] D. K. Campbell, J. F. Schonfeld, and C. A. Wingate, "Resonance Structure in Kink-Antikink Interactions in Φ^4 Theory," *Physica D* **9**, 1-32 (1983).

[8] M. Peyrard and D. K. Campbell, "Kink-Antikink Interactions in a Modified sine-Gordon Model," *Physica D* **9**, 33-51 (1983).

[9] D. K. Campbell and M. Peyrard, "Solitary Wave Collisions Revisited," *Physica D* **18**, 47-53 (1986).

[10] D. K. Campbell, M. Peyrard, and P. Sodano, "Kink-Antikink Interactions in the Double sine-Gordon Equation," *Physica D* **19**, 165-205 (1986).

[11] N. S. Manton and H. Merabet, "Φ^4 Kinks: Gradient Flow and Dynamics," *Nonlinearity* **10**, 3-18 (1997); G. W. Gibbons and N. S. Manton, "The Moduli Space Metric for Well-Separated BPS Monopoles," *Physics Letters B* **356**, 32-38 (1995).

[12] F. G. Mertens, H. J. Schnitzer, and A. R. Bishop, "Hierarchy of Equations of Motion for Nonlinear Coherent Excitations Applied to Magnetic Vortices," *Phys. Rev. B* **56** 5 (1997).

[13] H. Segur and M. D. Kruskal, "Nonexistence of small amplitude breather solutions in ϕ^4 field theory," *Phys. Rev. Lett.* **58** 747-750 (1987).

[14] B. Birnir, H. McKean, and A. Weinstein, "The Rigidity of Sine-Gordon Breathers," *Comm. Pure and App. Math.* **97**, 1043-1051 (1994).

[15] D. K. Campbell and M. Peyrard, "Chaos and Order in *Non*-Integrable Model Field Theories," pp. 305-334 in "*CHAOS/XAOC: Soviet-American Perspectives on Nonlinear Science*, D. K. Campbell, ed. (AIP Press, 1990).

[16] R. S. MacKay and S. Aubry, "Proof of Existence of Breathers for Time-Reversible or Hamiltonian Networks of Weakly Coupled Oscillators," *Nonlinearity* **7**, 1623-1643 (1994).

[17] S. Takeno, K. Kisoda, and A. J. Sievers, "Intrinsic Localized Vibrational Modes in Anharmonic Crystals," *Prog. Theor. Phys. Suppl.* **94** 242 (1988).

[18] S. Flach, "Existence of Localized Excitations in nonlinear Hamiltonian Lattices," *Phys. Rev. E.* **51**, 1503-1507 (1995).

[19] S. Flach and C. R. Willis, "Discrete Breathers," *Phys. Rep.* **295**, 181-264 (1998).

[20] X.-D. Wang, D. K. Campbell, and J.E. Gubernatis, "Quantum Double-Well Chain: Ground State Phases and Applications to Hydrogen-Bonded Materials," *Phys. Rev. B* **49**, 15485-15494 (1994).

FROM LANDAU'S HYDRODYNAMICAL MODEL TO FIELD THEORY MODELS OF MULTIPARTICLE PRODUCTION

FRED COOPER

Theoretical Division, Los Alamos National Laboratory,
Los Alamos, NM 87545
E-mail: fcooper@lanl.gov

We review the assumptions and domain of applicability of Landau's Hydrodynamical Model. By considering two models of particle production, pair production from strong electric fields and particle production in the linear σ model, we demonstrate that many of Landau's ideas are verified in explicit field theory calculations.

1 Introduction

In the early 1970's Peter Carruthers [1 2] (at the prodding of one of his graduate students Minh Duong Van) realized that Landau's Hydrodynamical model [3] explained the single particle rapidity distribution $dN/d\eta$ of pions produced in the inclusive reaction P-P $\rightarrow \pi$ + X extremely well.(see Fig. 1) He then asked me to think about the effect of adding thermal fluctuations and with the help of Graham Frye and Edmond Schonberg at Yeshiva University (and also some help from Mitchell Feigenbaum) we found the correct covariant method of adding thermal fluctuations to the hydrodynamic flow which is now called the Cooper-Frye-Schonberg formula [45]. Based on these ideas we were able to fit both the transverse momentum distribution as well as the rapidity distribution extremely well (see Fig. 2).

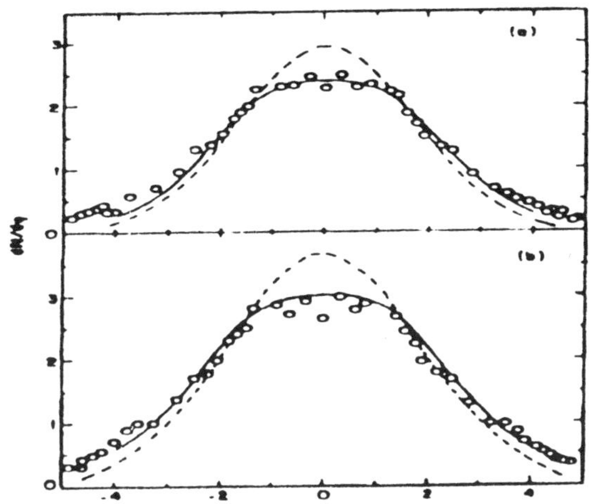

FIG. 1. Comparison of the experimental C.M. rapidity distribution of outgoing pions from Proton Proton collisions at the ISR at (a) $p_{ISR} = 15.4$ GeV/c and (b) $p_{ISR} = 26.7$ GeV/c compared with the no adjustable parameter result of the Landau Model from [2]

The hydrodynamical model, which was considered "heretical" when applied to proton-proton collisions,as opposed to Feynman scaling models, was resurrected by Bjorken [7] in 1983 to describe relativistic heavy ion collisions. The use of the hydrodynamical model to study heavy-ion collisions is now a sophisticated mini-industry using 3-dimensional fluid codes and sophisticated equations of state based on lattice QCD. The effects of resonance decays have also been recently included [8].

After the initial successes of the hydrodynamical model, Carruthers and Zachariasen [9 10] made the first transport approach to particle production based on the covariant Wigner-transport equation. This further work of Pete's inspired me and my collaborators David Sharp [11] and Mitchell Feigenbaum [12] to study particle production in $\lambda\phi^4$in a mean field approximation using the formalism espoused by Carruthers and Zachariasen. Our efforts in the 70's were hampered by two factors: first we were not sure what the appropriate field theory was, and secondly computers were not large enough or fast enough to perform accurate simulations of time evolution problems. This covariant transport theory approach turned out not to be the most convenient way to study initial value problems and would later be replaced by a non-covariant approach where the time was singled out.

Starting in the mid-80's two new approaches were taken to understand time evolution problems in field theory, both related in spirit to the covariant transport approach of Carruthers and Zachariasen. The first method was to assume a Gaussian ansatz in a time dependent variational method in the Schrodinger picture, [13 14]. The second method was to directly study the time evolution of the Green's functions in the Heisenberg picture in a leading order in large-N approximation [15 16 17]. Both these methods, which are mean-field approximations, lead to a well posed initial value problem for the time evolution of a field theory.

In the past few years, we have been able to consider two aspects of Relativistic Heavy Ion Collisions. The first aspect is connected with the production of the quark gluon plasma. The model we used was the popping of quark- antiquark pairs out of the vacuum due to the presence of Strong Chromoelectric fields. The mechanism we used was based on Schwinger's calculation of pair production from strong Electric Fields[20]. Using this model and assuming boost invariant kinematics we were able to show that many of Landau's assumptions were verified–namely that from the energy flow alone one could determine the particle spectrum, and that the fluid rapidity spectra was the same as the particle rapidity spectra when one is in a scaling regime. Also the hydrodynamical prediction for the dependence of the entropy density as a function of the proper time was verified at long times. Numerical evidence for this will be displayed below.

The second problem we considered was the dynamics of a non-equilibrium chiral phase transition. In this case because of the phase transition, one can obtain single particle spectra which are different from a local equilibrium flow such as that given by Landau's model. We found that when the evolution proceeds through the spinodal regime, low momentum modes grow exponentially for short periods of proper time. This leads to an enhancement of the low momentum spectrum over what would have been found in an equilibrium evolution. While considering this second problem we realized that the Cooper-Frye-Schonberg formula was valid even in a field theory evolution, provided one interprets the field theory interpolating number density as the single particle phase space distribution function of a classical transport theory [17].

It is safe to say that much of my career was stimulated by Pete encouraging me to understand multiparticle production in high energy collisions and by him freely sharing all of his intuition about this subject. It has taken 20 years to show that our original thinking back in 1973 was mostly correct!!

Landau's model [3] of multiparticle production was based on very few assumptions. He assumed that after a high energy collision some substantial fraction ($\approx 1/2$) of the kinetic energy in the center of mass frame was dumped into a Lorentz contracted disc with transverse size that of the smallest initial nuclei. He then assumed the flow of energy was describable by the relativistic hydrodynamics of a perfect fluid. The rest of the dynamics is embodied in the initial and final conditions.The initial condition on the energy density is that it is constant in a Lorentz contracted pancake. In our field theory calculations using strong fields, we will assume that all the energy density is in the initial semiclassical electric (chromoelectric) field. The final boundary condition is to state that when the energy density ε reduces to a critical value ε_c which is approximately one physical pion/ pion compton wavelength[3] in a comoving frame, then there are no more interactions and one then calculates the spectrum at that "freeze out" surface. If one then identifies the pion velocity with the collective velocity one equates the number of particles in a bin of particle rapidity with the energy in that bin divided by the energy of a single pion having that rapidity. This ansatz was later modified by myself and my collaborators by assuming that at breakup there was a local thermal distribution of pions in the comoving frame at temperature T_c described by

$$g(x,k) = g_\pi\{\exp[k^\mu u_\mu/T_c] - 1^{-1}. \tag{1}$$

The particle distribution was then given by:

$$E\frac{d^3N}{d^3k} = \frac{d^3N}{\pi dk_\perp^2 dy} = \int g(x,k)k^\mu d\sigma_\mu \tag{2}$$

where σ_μ is the surface defined by $\varepsilon = \varepsilon_c$. We have verified that both ideas are in agreement with our field theory calculations.

Let us now let us look at the hydrodynamics of a perfect fluid. In the rest frame (comoving frame) the stress tensor has the form:

$$T_{\mu\nu} = \text{diagonal}\ (\epsilon, p, p, p) \tag{3}$$

Boosting by the relativistic fluid velocity four vector $u^\mu(x,t)$ one has:

$$T_{\mu\nu} = (\varepsilon + p)u^\mu u^\nu - pg^{\mu\nu} \tag{4}$$

¿From a hydrodynamical point of view, flat rapidity distributions seen in multiparticle production in p-p as well as A-p and A-A collisions are a result of the hydrodynamics being in a scaling regime for the longitudinal flow. (More exact 3-D numerical simulations with sophisticated equations of state have now been performed. The interested reader can see for example[8]).

That is for $v = z/t$ (no size scale in the longitudinal dimension) the light cone variables τ, η:

$$z = \tau \sinh\eta; t = \tau\cosh\eta \tag{5}$$

become the fluid proper time $\tau = t(1 - v^2)^{1/2}$ and fluid rapidity:

$$\eta = 1/2 \ln[(t - x)/(t + x)] \Rightarrow 1/2 \ln[(1 - v)/(1 + v)] = \alpha \tag{6}$$

Letting $u^0 = \cosh\alpha; u^3 = \sinh\alpha$, we have when $v = zt$ that $\eta = \alpha$, the fluid rapidity. If one has an effective equation of state $p = p(\varepsilon)$ then one can formally define temperature and entropy as follows:

$$\varepsilon + p = Ts; d\varepsilon = Tds; \ln s = \int d\varepsilon/(\varepsilon + p) \tag{7}$$

Then the equation:

$$u^\mu \partial^\nu T_{\mu\nu} = 0$$

becomes:

$$\partial^\nu (s(\tau)u_\nu) = 0 \tag{8}$$

Which in 1 + 1 dimensions becomes

$$ds/d\tau + s/\tau = 0 \text{ or } s\tau = \text{constant} \tag{9}$$

Landau also assumed that the flow of energy is unaffected by the hadronization process and that the fluid rapidity can be identified in the out regime with particle rapidity. Thus after hadronization the number of pions found in a bin of fluid rapidity can be obtained from the energy in a bin of rapidity by dividing by the energy of a single pion having that rapidty. When the comoving energy density become of the order of ε_c we are in the out regime. This determines a surface defined by

$$\varepsilon_c(\tau_f) = m_\pi/V_\pi \tag{10}$$

On that surface of constant τ,

$$\begin{aligned}
\frac{dN}{d\eta} &= \frac{1}{m_\pi u^0}\frac{dE}{d\eta} = \frac{1}{m_\pi \cosh\alpha}\int T^{0\mu}\frac{d\sigma_\mu}{d\eta} \\
d\sigma_\mu &= A_\perp(dz, -dt) = 4\pi a^2 \tau_f(\cosh\eta, -\sinh\eta) \\
\frac{dN}{d\eta} &= \frac{A_\perp}{m_\pi \cosh\alpha}[(\varepsilon + p)\cosh\alpha\cosh(\eta - \alpha) - p\cosh\eta]
\end{aligned} \tag{11}$$

where $A_\perp$ is the transverse size of the system at freezeout. If we are in the scaling regime where $\eta = \alpha$ then

$$\frac{dN}{d\eta} = \frac{A_\perp}{m_\pi}\varepsilon(\tau_f)$$

which is a flat distribution in fluid rapidity. At finite energy, where scaling is not exact, only the central region is flat and instead one gets a distribution which is approximately Gaussian in rapidity. Exact numerical simulations (see for example [5]) show that the isoenergy curves do indeed follow a constant τ curve in the central region, so that the scaling result does apply for particle production in the central rapidity region at high but finite center of mass energy.

In our field theory approach, if we make the kinematical assumption that the quantum expectation values of measurables are solely a function of τ we will obtain a flat rapidity distribution for the distribution of particles. We also can prove, using a coordinate transformation, that the distribution of particles in fluid rapidity is exactly the same as the distribution of particles in particle rapidity. Furthermore we also will find that it is a good approximation to use eq. 11 to determine the spectra of particles. For the full single particle distribution $E\frac{dN}{d^3p}$ we find that the Cooper-Frye formula is valid with the identification of the interpolating number density with the single particle distribution function of classical transport theory.

2 Production and Time evolution of a quark-antiquark Plasma

Our model for the production of the quark-gluon plasma begins with the creation of a flux tube containing a strong color electric field. If the energy density of the chromoelectric field gets high enough (see below) the quark-anti quark pairs can be popped out of the vacuum by the Schwinger mechanism [20]. For simplicity, here we discuss pair production (such as electron-positron pairs) from an abelian Electric Field and the subsequent quantum back-reaction on the Electric Field. The physics of the problem can be understood for constant electric fields as a simple tunneling process. If the electric field can produce work of at least twice the rest mass of the pair in one compton wavelength, then the vacuum is unstable to tunnelling. This condition is:

$$eE\frac{\hbar}{mc} \geq 2mc^2$$

which leads to a critical electric field of order $\frac{2m^2c^3}{\hbar e}$

The problem of pair production from a constant Electric field (ignoring the back reaction) was studied by J. Schwinger in 1951 [20] . The WKB argument is as follows: One imagines an electron bound by a potential well of order $|V_0| \approx 2m$ and submitted to an additional electric potential eEx . The ionization probability is proportional to the WKB barrier penetration factor:

$$\exp[-2\int_o^{V_o/e} dx\{2m(V_o - |eE|x)\}^{1/2}] = \exp(-\frac{4}{3}m^2/|eE|)$$

The exact answer Schwinger found from calculating the Action was:

$$w = [\alpha E^2/(2\pi^2)]\sum_{n=1}^{\infty}\frac{(-1)^{n+1}}{n^2}\exp(-n\pi m^2/|eE|).$$

We are interested in going beyond Schwinger and determining both the production of pairs as well as the time evolution of the Electric field. We will directly solve the field equations in the large-N approximation as discussed in detail in [16]. We assume that the kinematics of ultrarelativistic high energy collisions results in a boost invariant dynamics so that all expectation values are functions of the proper time $\tau = \sqrt{t^2 - z^2}$.We introduce the light cone variables τ and η, which will be identified later with fluid proper time and rapidity . These coordinates are defined in terms of the ordinary lab-frame Minkowski time t and coordinate along the beam direction z by

$$z = \tau \sinh\eta \quad , \quad t = \tau\cosh\eta\,. \tag{12}$$

The Minkowski line element in these coordinates has the form

$$ds^2 = -d\tau^2 + dx^2 + dy^2 + \tau^2 d\eta^2\,. \tag{13}$$

Hence the metric tensor is given by

$$g_{\mu\nu} = \mathrm{diag}(-1, 1, 1, \tau^2). \tag{14}$$

The QED action in curvilinear coordinates is:

$$\begin{aligned} S &= \int d^{d+1}x\,(\det V)[-\frac{i}{2}\bar{\Psi}\tilde{\gamma}^\mu\nabla_\mu\Psi + \frac{i}{2}(\nabla_\mu^\dagger\bar{\Psi})\tilde{\gamma}^\mu\Psi \\ & -im\bar{\Psi}\Psi - \frac{1}{4}F_{\mu\nu}F^{\mu\nu}], \end{aligned} \tag{15}$$

where

$$\nabla_\mu\Psi \equiv (\partial_\mu + \Gamma_\mu - ieA_\mu)\Psi \tag{16}$$

Varying the action leads to the Heisenberg field equation:

$$(\tilde{\gamma}^\mu\nabla_\mu + m)\,\Psi = 0\,, \tag{17}$$

$$\left[\gamma^0\left(\partial_\tau + \frac{1}{2\tau}\right) + \gamma_\perp\cdot\partial_\perp + \frac{\gamma^3}{\tau}(\partial_\eta - ieA_\eta) + m\right]\Psi = 0\,, \tag{18}$$

and the Maxwell equation: $E = E_z(\tau) = -\dot{A}_\eta(\tau)$

$$\frac{1}{\tau}\frac{dE(\tau)}{d\tau} = \frac{e}{2}\left\langle\left[\bar{\Psi}, \tilde{\gamma}^\eta \Psi\right]\right\rangle = \frac{e}{2\tau}\left\langle\left[\Psi^\dagger, \gamma^0\gamma^3\Psi\right]\right\rangle. \tag{19}$$

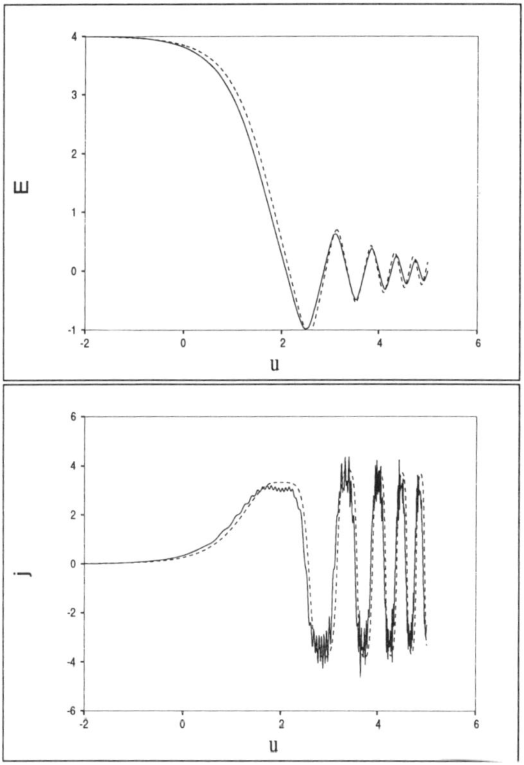

FIG. 4. Proper time evolution of E and j as a function of $u = \ln(\frac{\tau}{\tau_0})$ for an initial $E = 4$.

We expand the fermion field in terms of Fourier modes at fixed proper time: τ,

$$\begin{aligned}\Psi(x) &= \int [d\mathbf{k}] \sum_s [b_s(\mathbf{k})\psi^+_{\mathbf{k}s}(\tau)e^{ik\eta}e^{i\mathbf{p}\cdot\mathbf{x}} \\ &\quad + d^\dagger_s(-\mathbf{k})\psi^-_{-\mathbf{k}s}(\tau)e^{-ik\eta}e^{-i\mathbf{p}\cdot\mathbf{x}}].\end{aligned} \tag{20}$$

Letting:

$$\psi^{\pm}_{\mathbf{k}s} = \left[-\gamma^0 \left(\frac{d}{d\tau} + \frac{1}{2\tau}\right) - i\gamma_\perp \cdot \mathbf{k}_\perp - i\gamma^3 \pi_\eta + m\right] \chi_s \frac{f^{\pm}_{\mathbf{k}s}}{\sqrt{\tau}}. \tag{21}$$

$$\gamma^0 \gamma^3 \chi_s = \lambda_s \chi_s \tag{22}$$

with $\lambda_s = 1$ for $s = 1,2$ and $\lambda_s = -1$ for $s = 3,4$, we then get the mode equation:

$$\left(\frac{d^2}{d\tau^2} + \omega^2_{\mathbf{k}} - i\lambda_s \dot{\pi}_\eta\right) f^{\pm}_{\mathbf{k}s}(\tau) = 0, \tag{23}$$

$$\omega^2_{\mathbf{k}} = \pi^2_\eta + \mathbf{k}^2_\perp + m^2; \quad \pi_\eta = \frac{k_\eta - eA}{\tau}. \tag{24}$$

The back-reaction equation in terms of the modes is

$$\frac{1}{\tau}\frac{dE(\tau)}{d\tau} = -\frac{2e}{\tau^2} \sum_{s=1}^{4} \int [d\mathbf{k}](\mathbf{k}^2_\perp + m^2)\lambda_s |f^{+}_{\mathbf{k}s}|^2, \tag{25}$$

A typical proper time evolution of E and j is shown in fig. 4. Here an initial value of $E = 4$ was chosen.

2.1 Spectrum of Particles

To determine the number of particles produced one needs to introduce the adiabatic bases for the fields:

$$\begin{aligned} \Psi(x) &= \int [d\mathbf{k}] \sum_s [b^0_s(\mathbf{k};\tau) u_{\mathbf{k}s}(\tau) e^{-i\int \omega_{\mathbf{k}} d\tau} \\ &\quad + d^{(0)\dagger}_s(-\mathbf{k};\tau) v_{-\mathbf{k}s}(\tau) e^{i\int \omega_{\mathbf{k}} d\tau}] e^{i\mathbf{k}\cdot\mathbf{x}}. \end{aligned} \tag{26}$$

The operators $b_s(\mathbf{k})$ and $b^{(0)}_s(\mathbf{k};\tau)$ are related by a Bogolyubov transformation:

$$\begin{aligned} b^{(0)}_r(\mathbf{k};\tau) &= \sum \alpha^s_{\mathbf{k}r}(\tau) b_s(\mathbf{k}) + \beta^s_{\mathbf{k}r}(\tau) d^\dagger_s(-\mathbf{k}) \\ d^{(0)}_r(-\mathbf{k};\tau) &= \sum \beta^{*s}_{\mathbf{k}r}(\tau) b_s(\mathbf{k}) + \alpha^{*s}_{\mathbf{k}r}(\tau) d^\dagger_s(-\mathbf{k}) \end{aligned} \tag{27}$$

One finds that the interpolating phase space number density for the number of particles (or antiparticles) present per unit phase space volume at time τ is given by:

$$n(\mathbf{k};\tau) = \sum_{r=1,2} \langle 0_{in} | b^{(0)\dagger}_r(\mathbf{k};\tau) b^{(0)}_r(\mathbf{k};\tau) | 0_{in} \rangle = \sum_{s,r} |\beta^s_{\mathbf{k}r}(\tau)|^2 \tag{28}$$

This is an adiabatic invariant of the Hamiltonian dynamics governing the time evolution of the one and two point functions, and is therefore the logical choice as the particle number operator. The phase space distribution of particles (or antipartcles) in light cone variables is

$$n_{\mathbf{k}}(\tau) = f(k_\eta, k_\perp, \tau) = \frac{d^6 N}{\pi^2 dx_\perp^2 dk_\perp^2 d\eta dk_\eta}. \tag{29}$$

We introduce the particle rapidity y and $m_\perp = \sqrt{k_\perp^2 + m^2}$ defined by the particle 4-momentum in the center of mass coordinate system

$$k_\mu = (m_\perp \cosh y, k_\perp, m_\perp \sinh y) \tag{30}$$

The boost that takes one from the center of mass coordinates to the comoving frame is given by $\tanh\eta = v = z/t$, so that one can define the "fluid" 4-velocity in the center of mass frame as

$$u^\mu = (\cosh\eta, 0, 0, \sinh\eta) \tag{31}$$

We then find that the variable

$$\omega_k = \sqrt{m_\perp^2 + \frac{k_\eta^2}{\tau^2}} \equiv k^\mu u_\mu \tag{32}$$

has the meaning of the energy of the particle in the comoving frame. The variables

$$\tau = (t^2 - z^2)^{1/2} \qquad \eta = \frac{1}{2}\ln\left(\frac{t+z}{t-z}\right)$$

have as their canonical momenta

$$k_\tau = Et/\tau - k_z z/\tau = k^\mu u_\mu; \quad k_\eta = -Ez + tk_z \tag{33}$$

We change variables from (η, k_η) to (z, y) at a fixed τ where y is the particle rapidity defined by (30). We have

$$E\frac{d^3N}{d^3k} = \frac{d^3N}{\pi dy\, dk_\perp^2} = \int \pi dz\; dx_\perp^2\; J\; f(k_\eta, k_\perp, \tau) \tag{34}$$

where the Jacobian J is evaluated at a fixed proper time τ

$$J = \frac{m_\perp \cosh(\eta - y)}{\cosh\eta} = \frac{\partial k_\eta}{\partial z}|_\tau. \tag{35}$$

Calling the integration over the transverse dimension the effective transverse size of the colliding ions $A_\perp$ we then obtain that:

$$\frac{d^3N}{\pi dy\, dk_\perp^2} = A_\perp \int dk_\eta f(k_\eta, k_\perp, \tau) \equiv \frac{d^3N}{\pi d\eta\, dk_\perp^2} \tag{36}$$

This quantity is independent of y which is a consequence of the assumed boost invariance. Note that we have proven using the property of the Jacobean, that the distribution of particles in partical rapidity is the same as the distribution of particles in fluid rapidity!! verifying that in the boost-invariant regime that Landau's intuition was correct.

We now want to make contact with the Cooper- Frye Formalism. First we note that the interpolating number density depends on k_η and $k_\perp$ only through the combination:

$$\omega_k = \sqrt{m_\perp^2 + \frac{k_\eta^2}{\tau^2}} \equiv k^\mu u_\mu \tag{37}$$

Thus $f(k_\eta, k_\perp) = f(k_\mu u^\mu)$ and so it depends on exactly the same variable as the comoving thermal distribution! We also have that a constant τ surface (which is the freeze out surface of Landau) is parametrized as:

$$d\sigma^\mu = A_\perp(dz, 0, 0, dt) = A_\perp d\eta(\cosh\eta, 0, 0, \sinh\eta) \tag{38}$$

We therefore find

$$k^\mu d\sigma_\mu = A_\perp m_\perp \tau \cosh(\eta - y) = A_\perp |dk_\eta| \tag{39}$$

Thus we can rewrite our expression for the field theory particle spectra as

$$\frac{d^3N}{\pi dy\, dk_\perp^2} = A_\perp \int dk_\eta f(k_\eta, k_\perp, \tau) = \int f(k^\mu u_\mu, \tau) k^\mu d\sigma_\mu \tag{40}$$

where in the second integration we keep y and τ fixed. Thus with the replacement of the thermal single particle distribution by the interpolating number operator, we get via the coordinate transformation to the center of mass frame the Cooper-Frye formula.

Schwinger's pair production mechanism leads to an Energy Momentum tensor which is diagonal in the$(\tau, \eta, x_\perp)$ coordinate system which is thus a comoving one. In that system one has:

$$T^{\mu\nu} = \text{diagonal}\ \{\varepsilon(\tau), p_\parallel(\tau), p_\perp(\tau), p_\perp(\tau)\} \tag{41}$$

We thus find in this approximation that there are two separate pressures, one in the longitudinal direction and one in the transverse direction which is quite different from the thermal equilibrium case. Defining T, s via $\varepsilon + p_{\|} = Ts$, $d\varepsilon = Tds$, we have $\frac{d(\varepsilon\tau)}{d\tau} + p_{\|} = Ej_{\eta}$,

$$\frac{d(s\tau)}{d\tau} = \frac{Ej_{eta}}{T}$$

In the out regime we find $s\tau = \mathit{constant}$ as in the Landau Model, as is seen in fig.6.

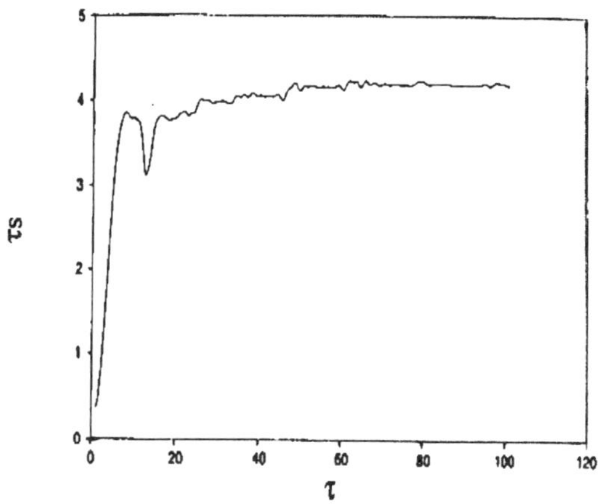

FIG. 6. Time evolution of $s\tau$ as a function of τ

For our one-dimensional boost invariant flow we find that the energy in a bin of fluid rapidity is just.

$$\frac{dE}{d\eta} = \int T^{0\mu} d\sigma_{\mu} = A_{\perp}\tau \cosh\eta\,\varepsilon(\tau) \tag{42}$$

which is just the (1 + 1) dimensional hydrodynamical result.Here however ε is obtained by solving the field theory equation rather than using an ultrarelativistic equation of state. This result does not depend on any assumptions of thermalization. We can ask if we can directly calculate the particle rapidity distribution from the ansatz:

$$\frac{dN}{d\eta} = \frac{1}{m\cosh\eta}\frac{dE}{d\eta} = \frac{A_{\perp}}{m}\varepsilon(\tau)\tau. \tag{43}$$

We see from fig. 8. that this works well even in our case where we have ignored rescattering, so that one does not have an equilibrium equation of state.

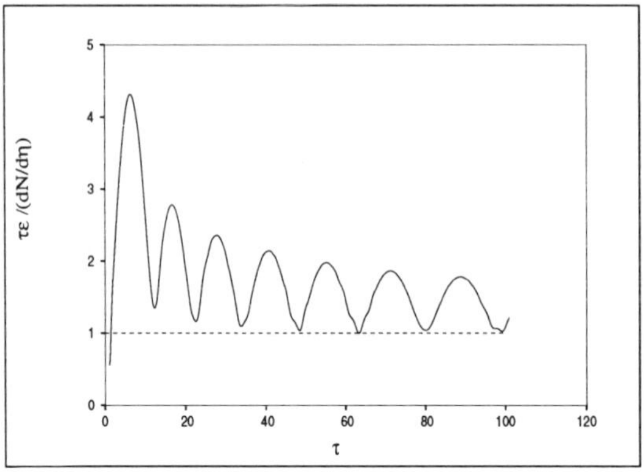

FIG. 8. The ratio of the approximate rapidity distribution $(\frac{E}{m\cosh y}\frac{dN}{dy})$ and exact rapidity distribution as a function of τ.

3 Dynamical evolution of a non-equilibrium Chiral Phase Transition

The model we use to discuss the chiral phase transition is the linear sigma model described by the Lagrangian:

$$L = \frac{1}{2}\partial\Phi\cdot\partial\Phi - \frac{1}{4}\lambda(\Phi\cdot\Phi - v^2)^2 + H\sigma. \tag{44}$$

The mesons form an $O(4)$ vector $\Phi = (\sigma, \pi_i)$ This can be written in an alternative form by introducing the composite field: $\chi = \lambda(\Phi\cdot\Phi - v^2)$.

$$L_2 = -\frac{1}{2}\phi_i(\Box + \chi)\phi_i + \frac{\chi^2}{4\lambda} + \frac{1}{2}\chi v^2 + H\sigma \tag{45}$$

The effective action to leading order in large N is given by [26]

$$\Gamma[\Phi,\chi] = \int d^4x[L_2(\Phi,\chi,H) + \frac{i}{2}N\mathrm{tr}\ \ln G_0^{-1}] \tag{46}$$

$$G_0^{-1}(x,y) = i[\Box + \chi(x)]\ \delta^4(x-y)$$

Varying the action we obtain:

$$[\Box + \chi(x)]\pi_i = 0 \qquad [\Box + \chi(x)]\sigma = H, \tag{47}$$

where here and in what follows, π_i,σ and χ refer to expectation values. Varying the action we obtain

$$\chi = -\lambda v^2 + \lambda(\sigma^2 + \pi \cdot \pi) + \lambda N G_0(x,x). \tag{48}$$

We again assume boost invariant kinematics [5 7]. We then get the equations:

$$\begin{aligned} &\tau^{-1}\partial_\tau\ \tau\partial_\tau\ \Phi_i(\tau) + \ \chi(\tau)\ \Phi_i(\tau) = H\delta_{i1} \\ &\chi(\tau) = \lambda\big(-v^2 + \Phi_i^2(\tau) + N G_0(x,x;\tau,\tau)\big), \end{aligned} \tag{49}$$

To determine the Green's function $G_0(x,y;\tau,\tau')$ we introduce the auxiliary quantum field $\phi(x,\tau)$ which obeys the sourceless equation:

$$\Big(\tau^{-1}\partial_\tau\ \tau\partial_\tau\ - \tau^{-2}\partial_\eta^2 - \partial_\perp^2 + \chi(x)\Big)\phi(x,\tau) = 0. \tag{50}$$

$$G_0(x,y;\tau,\tau') \equiv < T\{\phi(x,\tau)\ \phi(y,\tau')\} > .$$

We expand the quantum fields in an orthonormal basis:

$$\phi(\eta, x_\perp, \tau) \equiv \frac{1}{\tau^{1/2}} \int [d^3\mathrm{k}]\big(\exp(ikx) f_\mathrm{k}(\tau)\ a_\mathrm{k}\ + h.c.\big)$$

where $kx \equiv k_\eta \eta + \vec{k}_\perp \vec{x}_\perp$, $[d^3\mathrm{k}] \equiv dk_\eta d^2k_\perp/(2\pi)^3$. The mode functions and χ obey:

$$\ddot{f}_\mathrm{k} + (\frac{k_\eta^2}{\tau^2} + \vec{k}_\perp^2 + \chi(\tau) + \frac{1}{4\tau^2}) f_\mathrm{k} = 0. \tag{51}$$

$$\chi(\tau) = \lambda\Big(-v^2 + \Phi_i^2(\tau) + \frac{1}{\tau} N \int [d^3\mathrm{k}] |f_\mathrm{k}(\tau)|^2\ (1 + 2\ n_\mathrm{k})\Big). \tag{52}$$

We notice that when χ goes negative, the low momentum modes with

$$\frac{k_\eta^2 + 1/4}{\tau^2} + \vec{k}_\perp^2 < |\chi|$$

grow exponentially. However these modes then feed back into the χ equation and this exponential growth then gets damped. It is these growing modes that lead to the possiblity of growing domains of DCC's as well as a modification of the low momentum distribution of particles from a thermal one. The parameters of the model are fixed by physical data. In the vacuum state $\chi_0\sigma_0 = m_\pi^2\sigma_0 = H$, so that $\sigma_0 = f_\pi$= 92.5 MeV. The vacuum gap equation is

$$m_\pi^2 = -\lambda v^2 + \lambda f_\pi^2 + \lambda N \int_0^\Lambda [d^3\mathrm{k}] \frac{1}{2\sqrt{k^2 + m_\pi^2}}.$$

This leads to the mass renormalized gap equation:

$$\chi(\tau) - m_\pi^2 = -\lambda f_\pi^2 + \lambda \Phi_i^2(\tau)$$
$$+\frac{\lambda}{\tau} N \int [d^3\mathrm{k}] \{ |f_\mathrm{k}(\tau)|^2 \ (1 + 2\ n_\mathrm{k}) - \frac{1}{2\sqrt{k^2+m^2}} \}. \tag{53}$$

λ is chosen to fit low energy scattering data as discussed in [17].

We assume that the inititially (at $\tau_0 = 1$) the system is in local thermal equilibrium in a comoving frame we have

$$n_k = \frac{1}{e^{\beta_0 E_k^0} - 1} \tag{54}$$

where $\beta_0 = 1/T_0$ and $E_k^0 = \sqrt{\frac{k_\eta^2}{\tau_0^2} + \vec{k}_\perp^2 + \chi(\tau_0)}$.

The initial value of χ is determined by the equilibrium gap equation for an initial temperature of $200MeV$ and is $.7fm^{-2}$ and the initial value of σ is just $\frac{H}{\chi_0}$. The phase transition in this model occurs at a critical temperature of

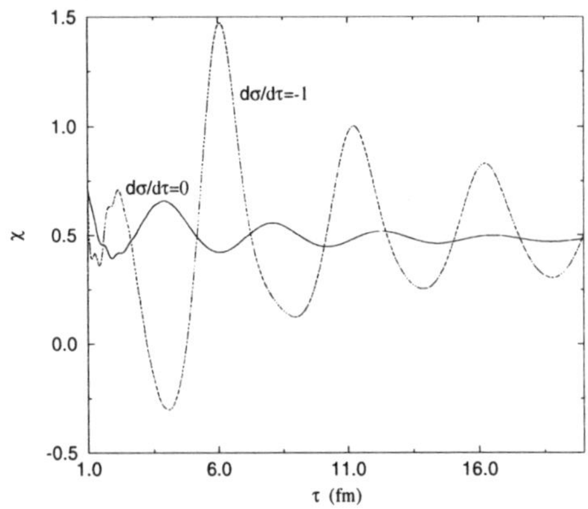

$160MeV$.

FIG. 10.Proper time evolution of the χ field for two different initial values of $\dot{\sigma}$.

Fig.10 displays the results of the numerical simulation for the evolution of χ (51)–(52). We display the auxiliary field χ in units of fm^{-2} , the classical fields Φ in units of fm^{-1} and the proper time in units of fm $(1fm^{-1} = 197MeV)$ for two simulations, one with an instability $(\dot{\sigma}|_{\tau_0} = -1)$ and one without $(\dot{\sigma}|_{\tau_0} = 0)$.

We notice that for both initial conditions, the system eventually settles down to the broken symmetry vacuum result as a result of the expansion. To determine the spectrum of particles we again introduce the interpolating number density which is defined by expanding the fields in terms of mode functions f_k^0 which are first order in an adiabatic expansion of the mode equation.

In terms of the initial distribution of particles $n_0(k)$ and β we have:

$$\begin{aligned} n_k(\tau) &\equiv f(k_\eta, k_\perp, \tau) = < a_k^\dagger(\tau) a_k(\tau) > \\ &= n_0(k) + |\beta(k,\tau)|^2(1 + 2n_0(k)). \end{aligned} \tag{55}$$

where $\beta(k,\tau) = i(f_k^0 \frac{\partial f_k}{\partial \tau} - \frac{\partial f_k^0}{\partial \tau} f_k)$, $n_k(\tau)$ is the interpolating number density. The distribution of particles is

$$f(k_\eta, k_\perp, \tau) = \frac{d^6 N}{\pi^2 dx_\perp^2\, dk_\perp^2\, d\eta dk_\eta}. \tag{56}$$

Changing variables from (η, k_η) to (z, y) at a fixed τ we have

$$\begin{aligned} E\frac{d^3N}{d^3k} &= \frac{d^3N}{\pi dy\, dk_\perp^2} = \int \pi dz\; dx_\perp^2\;\; J\; f(k_\eta, k_\perp, \tau) \\ &= A_\perp \int dk_\eta f(k_\eta, k_\perp, \tau) = \int f(k_\eta, k_\perp, \tau) k^\mu d\sigma_\mu. \end{aligned} \tag{57}$$

We wish to compare our nonequilibrium calculation with the results of the hydrodynamical model in the same boost-invariant approximation. In the hydrodynamical model of heavy ion collisions [5], the final spectra of pions is given by a combination of the fluid flow and a local thermal equilibrium distribution in the comoving frame.

$$E\frac{d^3N}{d^3k} = \frac{d^3N}{\pi dk_\perp^2\, dy} = \int g(x,k) k^\mu d\sigma_\mu \tag{58}$$

Here $g(x,k) = g_\pi \{\exp[k^\mu u_\mu / T_c] - 1\}^{-1}$ is the single particle relativistic phase space distribution function for a thermal distribution. 690 In Figures 12 and 13 we compare the boost invariant hydrodynamical result for the transverse momentum distribution using critical temperatures of $T_c = 140, 200$ MeV to the two nonequilibrium cases represented in figure 10. Figure 12 pertains to the initial condition $\dot{\sigma}|_{\tau_0} = -1$ In this case there is a regime where the effective mass becomes negative and we see a noticable enhancement of the low transverse momentum spectra. We have normalized both results to give the same total center of mass energy E_{cm}. Figure 13 corresponds to the initial condition $\dot{\sigma}|_{\tau_0} = 0$. Here we notice that there is a little enhancement at low transverse momenta.

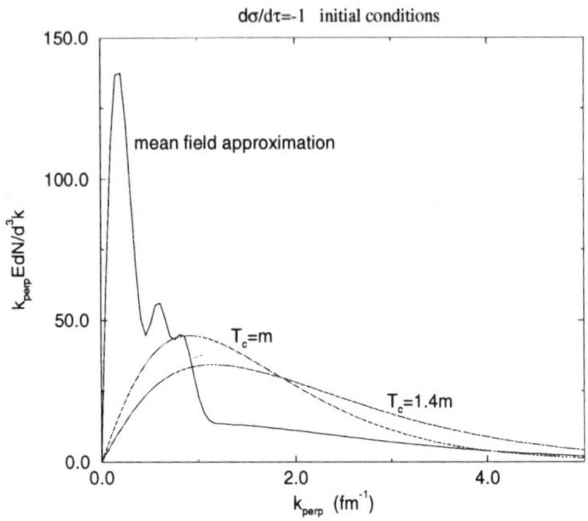

FIG. 12. Single particle transverse momentum distribution for $\dot{\sigma} = -1$ initial conditions compared to a local equilibrium Hydrodynamical calculation with boost invariance.

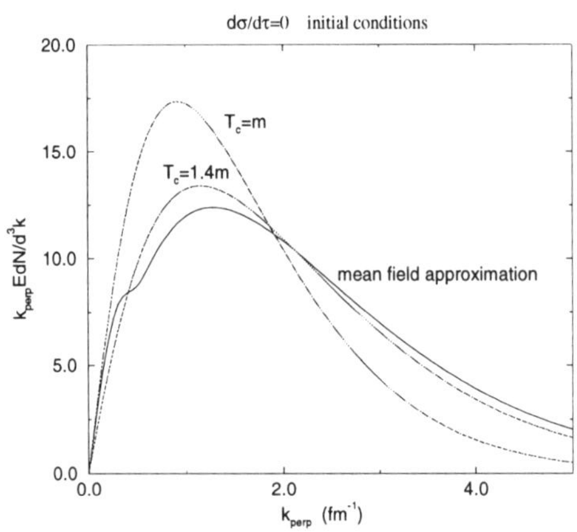

FIG. 13. Single particle transverse momentum distribution for $\dot{\sigma} = 0$ initial conditions compared to a local equilibrium Hydrodynamical calculation with boost invariance.

So we see that a non-equilibrium phase transition taking place during a time evolving quark-gluon or hadronic plasma can lead to an enhancement of the low momentum transverse momentum distribution. In particular, if a Centauro type event is not accompanied by such an enhancement one would be suspicious of ascribing this event to the production of disoriented chiral condensates as a result of a rapid quench. For the case when we have massless goldstone pions in the σ model ($H = 0$)[27],then χ goes to zero at large times. In that case we find the amusing fact that the equation of state becomes $p = \varepsilon/3$ at late times. This is true even though the final particle spectrum is far from thermal equilibrium. This equation of state is shown in Fig 14.

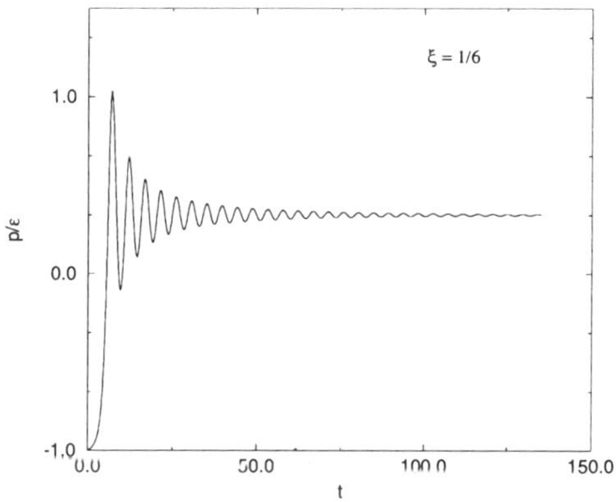

FIG. 14. Equation of state $\frac{p}{\varepsilon}$ as a function of τ for the massless σ model where we start from a quench.

4 ACKNOWLEDGEMENTS

The work presented here was done in collaboration with Emil Mottola, Salman Habib, Yuval Kluger, Juan Pablo Paz, Ben Svetitsky, Judah Eisenberg, Paul Anderson, So-Young Pi, John Dawson, David Sharp, Mitchell Feigenbaum, Edmond Schonberg and Graham Frye. This work was supported by the Department of Energy.

1. P. Carruthers and D. V. Minh, Phys. Lett. **42B**, 597 (1973).
2. P. Carruthers, Ann. N.Y. Acad. Sci **229**, 91 (1974).
3. L.D. Landau, Izv. Akad. Nauk SSSR **17**, 51 (1953).
4. F. Cooper and G. Frye, Phys. Rev. **D10**, 186 (1974).

5. F. Cooper, G. Frye and E. Schonberg Phys. Rev. **D11**, 192 (1975).
6. F. Cooper and E. Schonberg, Phys. Rev. Lett. **30**, 880-883 (1973).
7. J.D. Bjorken, Phys. Rev. **D27** 140, 1983
8. B.R. Schlei, U. Ornik, M. Plumer, D. Strottman, R.M. Weiner, Phys. Lett. **B 376**, 212 (1996);Phys. Rev. **C54**.1381 (1996).
9. P. Carruthers and F. Zachariasen 1974 APS/DPF conference proceedings. AIP Conference Proceedings no. 23 (1975) 481.
10. P. Carruthers and F. Zachariasen, Phys. Rev. **D13** 950 (1976)
11. F. Cooper and D.H. Sharp, Phys. Rev. **12**, 1123 (1975).
12. F. Cooper and M. Feigenbaum, Phys. Rev. **14**, 583 (1976).
13. F. Cooper, S. Y. Pi, and P. Stancioff, Phys. Rev. D**34**, 3831 (1986).
14. S.Y. Pi and Samiullah Phys. Rev. **D36**, 3128 (1987)
15. F. Cooper and E. Mottola, Phys. Rev. D**36**, 3114 (1987).
16. F. Cooper, J.M. Eisenberg, Y. Kluger, E. Mottola, and B. Svetitsky, Phys. Rev. D 48 (1993) 190.
17. F. Cooper, Y. Kluger, E. Mottola and J.P. Paz Phys. Rev.**D51**,2377 (1995)hep-th/9404357
18. F. Cooper and E. Mottola, Phys. Rev. D **40**, 456 (1989)
19. F.Cooper,E. Mottola, B.Rogers and P. Anderson, Proceedings of the Santa Fe Workshop on 'Intermittency in High-Energy Collisions', March 1990 (World Scientific). Eds. F. Cooper, R. Hwa and I. Sarcevic (1991).
20. J. Schwinger, Phys. Rev. Lett. **82**,664 (1951)
21. S. Weinberg, *Gravitation and Cosmology* (Wiley,New York, 1972).
22. A. Anselm, Phys. Lett. **B217**, 169 (1989); A. Anselm and M. Ryskin, Phys. Lett. **B226**, 482 (1991).
23. J.D. Bjorken, Int. Jour. of Mod. Phys.**A7**, 4189 (1992); Acta Phys. Pol. **B23** 561 (1992).
24. K. Rajagopal and F. Wilczek, Nucl. Phys. **B379**, 395 (1993); S. Gavin, A. Gocksch and R.D. Pisarski, Phys. Rev. Lett. **72**,2143 (1994)
25. C.M.G. Lattes, Y. Fujimoto and S. Hasegawa, Phys. Rep. **65**, 151 (1980).
26. F. Cooper, Y.Kluger, E.Mottola, J.P. Paz, Phys. Rev. **D51**,2377 (1995).
27. F. Cooper, S. Habib,Y. Kluger and E. Mottola "Nonequilibrium Dynamics of Symmetry Breaking in $\lambda\phi^4$ Field Theory" hep-ph/96 10345

WAVELET ANALYSIS OF A DISORIENTED CHIRAL CONDENSATE

I. SARCEVIC[†], Z. HUANG[†], R. THEWS[†] and X.-N. WANG[‡]
[†]*Department of Physics, University of Arizona, Tucson, AZ 85721*
[‡]*Nuclear Science Division, LBNL, Berkeley, CA 94720*

We present a novel method for studying the formation of a disoriented chiral condensate (DCC) in high energy heavy-ion collisions utilizing a discrete wavelet transformation. We show that the probability distributions of the neutral pion fraction for various rapidity-bin sizes have distinctive shapes in the case of a DCC and exhibit a delay in approaching the Gaussian distribution required by the Central Limit Theorem. We find the wavelet power spectrum for a DCC to have a strong dependence on the scale while an equilibrium system and the standard dynamical models such as HIJING have a flat spectrum. We demonstrate the applicability of wavelet analysis on the JACEE 4LIIG-27 cosmic ray event.

1 Introduction

In heavy-ion collisions there is a possibility of supercooling of the interaction region after the collision resulting in a formation of "unconventionally" oriented vacuum configurations, the so-called Disoriented Chiral Condensate (DCC).[1] In the language of the linear σ-model, the produced vacuum is not found in the σ direction but aligned to some general orientation in $O(4)$ space. High energy heavy-ion collisions may provide a suitable condition for such a phenomenon to occur: the produced energy density is high enough so that the chiral symmetry may be initially restored and the longitudinal expansion is rapid enough to lead to the supercooling. Detection of this interesting phenomenon would provide valuable information on the vacuum structure of the strong interaction and the nature of the chiral phase transition. However, in case of heavy-ion collisions many uncorrelated small domains of DCC may be formed with different orientations so that the distinctive signature of the single disoriented condensate may be largely smeared. The integrated probability distribution of the neutral pion fraction f emitted from a disoriented region, predicted to be $P(f) = 1/2\sqrt{f}$,[1] would then become Gaussian, a generic distribution. However, we find that a new method based on the wavelet transformation can reveal the formation of DCC in heavy-ion collisions. In particular, the wavelet transformed probability distribution of the neutral pion fraction and the power spectrum exhibit a strong dependence on the scale characterized by the size in rapidity.

2 Dynamical Model of DCC

The dynamical process of chiral phase transition in QCD is presently not completely understood. If there is significant supercooling before the plasma hadronizes, the evolution of the chiral order parameter is likely to be out of equilibrium. The linear σ-model is shown to characterize the basic feature of the chiral phase transition since it is in the same universality class as QCD. It also gives a good description of soft pion dynamics such as pion-pion scattering which dominates the late stages of evolution of the system. In the linear σ-model, the dynamical field Φ consists of four components $(\sigma, \pi_1, \pi_2, \pi_3)$ denoting the sigma and pion fields. In high multiplicity events, where large numbers of soft pions are produced in high-energy collisions, one attempts to describe the system by the classical field theory, because the quantum effects are expected to be small for states with large occupation numbers. The dynamical evolution of the interacting pion system is governed by the classical equation of motion derived from the linear σ-model. The initial condition can be chosen to represent the possible configurations before the phase transition.

We simulate an ensemble of random fields distributed in space with a geometry consistent with the longitudinal expansion. The field distributions are generated by general Gaussian random distributions with a mean field $\langle\Phi\rangle$ and a variance $\langle\delta\Phi^2\rangle$ which determine whether the system is in or out of equilibrium initially. The mean field which characterizes the order parameter evolves according to the equation of motion in a Hartree approximation

$$\left(\frac{\partial^2}{\partial t^2} - \Delta^2\right)\langle\Phi\rangle = \frac{\partial V_{\rm eff}(\langle\Phi\rangle, \langle\delta\Phi^2\rangle)}{\partial\langle\Phi\rangle}; \tag{1}$$

where the effective potential for the mean field $\langle\Phi\rangle$, $V_{\rm eff}(\langle\Phi\rangle, \langle\delta\Phi^2\rangle)$ depends on the mean field and the fluctuation $\langle\delta\Phi^2\rangle$ and is given by[2]

$$V_{\rm eff}(\langle\Phi\rangle, \langle\delta\Phi^2\rangle) = \frac{\lambda}{4}(\langle\Phi\rangle^2 + \langle\delta\Phi^2\rangle - v^2)^2 + H\langle\sigma\rangle. \tag{2}$$

Parameters λ, v and H are determined from the low energy data.[3] The fluctuation $\langle\delta\Phi^2\rangle$ can be thought of as the "effective temperature" though the temperature is not well defined when the system is out of equilibrium. In the supercooling case, the fluctuation is small (the "effective temperature" is low) so that $V_{\rm eff}$ exhibits the shape of a Mexican hat while the mean field lies on the top of the "hat". As the system evolves, the mean field rolls down from the chirally symmetric false vacuum to the true chirally asymmetric vacuum. For low momentum modes, such a rolling down is not stable. As a result, some domains of correlated pions begin to form due to this instability. This scenario is referred to as the quenching mechanism.[3] An equilibrium evolution is a situation where the initial mean field is given by the equilibrium position of the effective potential at certain "effective temperature". In such a case, there is no initial instability. Whether or not the domain structure can develop in such a case is determined by the subsequent dynamical evolution according to (1).

3 Wavelets

The novel method that we propose for studying the formation of DCC is a multi-resolution analysis performed by a discrete wavelet transformation (DWT)[4] which has been found effective in systematically detecting structures of various scales in various physics problems. For any one-dimensional sample of a point distribution $f(x)$, such as the rapidity distribution of the neutral pion fraction in the interval $[0, 1]$ with resolution Δx, can be represented as a histogram of 2^J bins where $J = \mathrm{mod}(|\ln\Delta x|/\ln 2) + 1$

$$f(x) \equiv f^{(J)}(x) = \sum_{k=0}^{2^J-1} f_{Jk}\phi_{Jk}(x), \tag{3}$$

where $\phi_{Jk}(x)$ are the piecewise constant functions called the mother functions. The histogram function at the next finer resolution is the average of adjacent bins at the finest scale. Evidently some detail is lost compared to (3). The difference between (3) and $f^{(J-1)}$ can be fully expressed in terms of the difference functions $\psi_{J-1,k}(x)$ which are stairwell functions called father functions. One defines the difference amplitudes or the DWT coefficients

$$\tilde{f}_{jk} = 2^j \int f(x)\psi_{jk}(x)dx, \tag{4}$$

where the basis functions $\psi_{jk}(x)$ have a unique property: they are related to each other by scaling and translation: $\psi_{jk}(x) = \psi(2^j x - k)$. The index j denotes the resolution scale and k the position at each scale. The idea of multiresolution analysis is to find representations of the sample function $f(x)$ at various scales.

The main idea of using the wavelet analysis is to attempt an unbiased separation of the DCC physics from the various "backgrounds" which cannot be otherwise easily eliminated. The usual correlation method is not so effective here because the different scale correlations are mixed together in the same phase space. The discrete wavelet transformation (DWT), on the other hand, has an extra dimension – the scale which can be used to separate these correlations. The main source of the "backgrounds" are the fluctuations at small correlation scale due to pion radiation from the uncorrelated regions in space-time, which can coexist with a DCC domain. The background can also come from the partial thermalization of the system leading to thermal correlations.

4 Probability Distribution of the Neutral-Pion Fraction

We have simulated the evolution of the pion fields using a classical linear σ-model. The initial condition is chosen to model quenched scenario.[3] After evolving this nonequilibrium state from $\tau_{\text{initial}} = 1$ to $\tau_{\text{final}} = 5$ fm/c, we form a one-dimensional data array of

$$f(\eta) = \frac{\pi_3^2(\eta)}{\pi_1^2(\eta) + \pi_2^2(\eta) + \pi_3^2(\eta)}, \tag{5}$$

for each event as the spatial rapidity (η) distribution of the π_3-fraction f, where π_i $(i = 1, 2, 3)$ are the three Cartesian components of the classical pion field. The lattice spacing in η is chosen to be $\Delta\eta = 0.08$. Thus initially the pion fields are automatically correlated within $\Delta\eta$ but completely random at scales beyond $\Delta\eta$. The maximal rapidity interval is chosen to be $[-5, 5]$, with a periodic boundary condition. The total number of bins is $2\eta_{\text{max}}/\Delta\eta = 2^7$, so $J = 7$ is the finest scale. We have calculated the father-function representation of DCC samples and random-noise sample using D4-wavelet. We have considered location independent quantities such as the probability distribution of father-function coefficients (FFC). At a given scale, the FFC's are not aware of the existence of the clustering structures at the subsequent scales – they only measure the fluctuations with that "frequency" $1/\Delta\eta$. The distributions should not look very different as j decreases toward j_d at which clustering occurs, as the FFC's only record the "noise". They should be quite similar to those of the random-noise samples except that the "noise levels" are lower in the DCC samples which implies a smoother rapidity distribution. Thus, they should be more peaked around $\langle\tilde{f}\rangle \simeq 0$. As j decreases further the clusters begin to appear, and fluctuations will grow as the FFC's measure the difference between the structure and the structure-less regions. Thus the FFC's at different scales can be thought of as measuring the fluctuations using various sized "rulers". When the size of a ruler matches the size of a structure, the strength of fluctuations grows and thus the probability distribution of FFC's becomes more spread-out when $j \le j_d$. The above picture is supported in our demonstration as shown in Fig. 1 where the probability distributions of FFC's at each scale for both the DCC samples and the noise samples are plotted. For the noise samples as one may anticipate, the probability histograms look alike at all scales, indicating the nonexistence of any scale structures. We define j_d

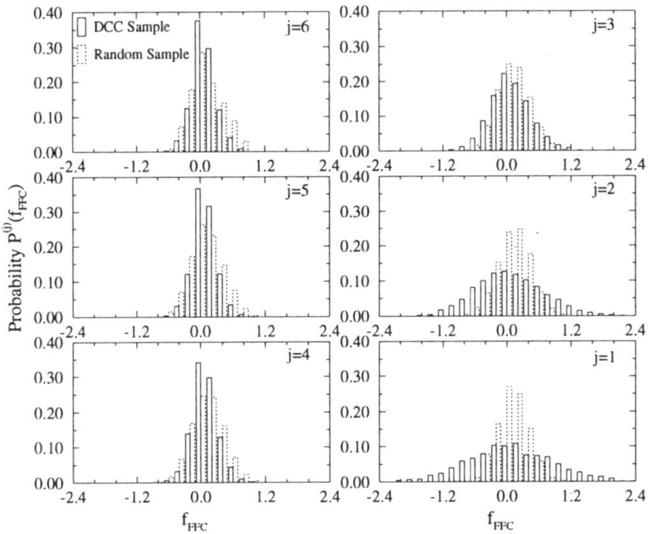

Figure 1: The probability distributions of the FFC's for both DCC samples and random samples at each scale j.

to be the scale for the DCC clustering, beyond which the fluctuations surpass the pure random-noise fluctuations. Clearly from Figure 1 this happens at a scale between $j = 4$ and $j = 3$.

5 Wavelet Power Spectrum

Another way to analyze the fluctuations at each scale is the wavelet power spectrum. Similar to the Fourier analysis, a Parseval Theorem for DWT utilizes the complete and orthonormal properties of the DWT:

$$\frac{1}{L}\int_0^L |f(x)|^2 dx = \frac{1}{L}\sum_{j=0}^{\infty}\frac{1}{2^j}\sum_{k=0}^{2^j-1}|\tilde{f}_{jk}|^2 , \tag{6}$$

where $\tilde{f}_{jk}$ are the FFC's. Thus, one can define the power spectrum with respect to the wavelet basis at each j as

$$P_j = \frac{1}{2^j}\sum_{k=0}^{2^j-1}|\tilde{f}_{jk}|^2 . \tag{7}$$

P_j is then the power of fluctuations on length scale $L/2^j$. It is also the mean square of the probability distributions of FFC's in Figure 1. The random-noise

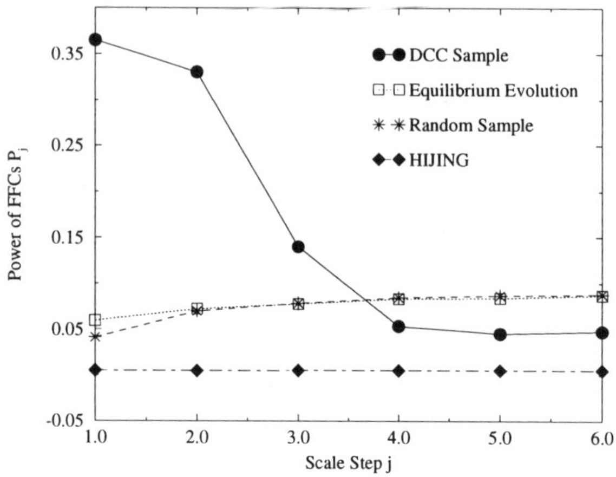

Figure 2: The power spectra of the FFC's for different dynamical scenarios.

sample features a flat power spectrum, i.e. the power of fluctuations is the same at any scale. For the DCC sample there should be a flat spectrum when $j \geq j_d$ and some power build-up should show up when $j \leq j_d$. We demonstrate the power spectrum for the rapidity distribution of the FFC's for three different dynamical scenarios in Figure 2. Indeed, a plateau structure is evident in Fig. 2 as j goes down from 6 to 4 in the DCC power spectrum. The power level of that plateau is lower than those in the annealing equilibrium evolution scenario and the random-noise sample which indicates a smoother distribution within a DCC domain. The power starts to build up as j goes down below 4 and rises rather quickly, departing distinctively from the annealing and the random-noise cases. The crossing point in Fig. 2 is found to be at $j_d = 3.6$ which unambiguously suggests the existence of the DCC clustering with a typical size of $\Delta\eta_d = 2\eta_{\max}/2^{j_d} \simeq 0.8$ units in rapidity. Also plotted is the power spectrum from HIJING Monte Carlo data which is also flat, consistent with the random-noise case.

The power spectrum can be thought of as describing the scale dependence of the "temperature". The annealing scenario (following an equilibrium evolution) or the noise sample features a uniformly flat spectrum and thus possesses a universal (global) temperature; while the rapid rise of the spectrum in the DCC case reflects the highly nonequilibrium nature of the fluctuations and the sensitivity of the temperature on the scale change.

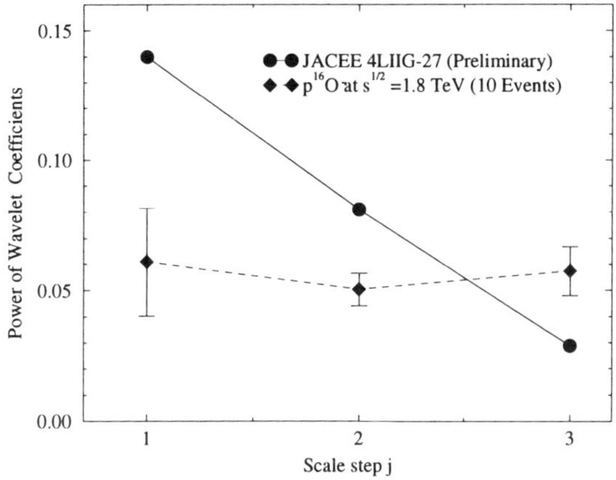

Figure 3: The power spectra of the JACEE event and 10 events generated by the HIJING Monte Carlo program.

6 Applicability to Experimental Data

Our method is directly applicable to experimental data as long as there is a sufficiently large number of particles, say, about 10 at the finest resolution. When the multiplicity is small, the neutral-pion-fraction distribution depends very sensitively on the multiplicity and its fluctuation in each bin. In this case, the neutral-pion fraction can only take a few discrete values and its probability distribution is largely controlled by the multiplicity distribution. This consideration sets the limitation for the applicability of our method. For the experimental rapidity resolution of $\Delta y \sim 0.1$, for example, the average multiplicity of $dN/dy \sim 100$ or larger is needed for the optimal use of the DWT. Such a large multiplicity is more likely in heavy-ion collisions than in hadronic collisions. Although there are no conceptual difficulties in applying our method to large multiplicity events in hadronic colliders such as LHC, the DWT is best suitable to discern the disoriented chiral condensate in high energy heavy-ion collisions at CERN/SPS, BNL/RHIC and CERN/LHC. In principle, our method is also applicable to a few events with large multiplicity such as cosmic ray experiments.

As an illustration of the applicability of our method we have done a wavelet analysis on the JACEE event[5] which contains unusually large number of neutral pions relative to the number of charged pions. This particular event is most likely initiated by a proton primary with energy of about 2 TeV. We divide the

JACEE phase space in rapidity into 2^4 bins and compute the neutral fraction $f = N_\gamma/2N_{\rm tot}$ in each rapidity bin. The calculated wavelet power spectrum is plotted in Fig. 3. We also use HIJING Monte Carlo to generate minimum bias events whose multiplicity coincides with that of JACEE event. The power spectrum of the Monte Carlo events is also plotted in Fig. 3. The JACEE event has a rising spectrum as j decreases, showing the existence of the cluster structure, while the Monte Carlo events have a flat spectrum. One should keep in mind, however, that the experimental analysis of this particular event has been particularly challenging due to the difficulty to deconvolve the primary interaction features from the secondary cascades observed.

7 Conclusions

A disoriented chiral condensate is a very interesting phenomenon of misaligned vacuum possibly created in high-energy heavy-ion collisions. Discovery of DCC requires a substantial suppression of background coming from ordinary QCD hard processes and statistical fluctuations. The discrete wavelet transformation is a very efficient and reliable method for separating the background from the signal according to their different physical scales. Wavelet analysis opens a new window for studying the scale dependence of various features of clustering structure present in the DCC. We conclude that the signal for a disoriented chiral condensate in heavy-ion collisions may be discovered by observing a delayed approach of the probability distribution of the neutral-pion fraction to the Gaussian distribution (as predicted by the central limit theorem) and by a distinctive increase of the wavelet power spectrum for large rapidity bins.

Acknowledgments

This work was supported in part by the DOE grants DE-FG03-93ER40792 and DE-FG02-85ER40213.

References

1. J. -P. Blaizot and A. Krzywcki, *Phys. Rev.* D**46**, 246 (1992); J. D. Bjorken *et al.*, in *Les Recontres de la Vallée D'Aoste, La Thuile*, ed. M. Greco (Editions Frontier, 1993) p. 507.
2. Z. Huang and X.-N. Wang, *Phys. Rev.* D**49**, R4339 (1994).
3. K. Rajagopal and F. Wilczek, *Nucl. Phys.* B**404**, 577 (1993).
4. Z. Huang, I. Sarcevic, R. Thews and X.-N. Wang, *Phys. Rev.* D**54**, 750 (1996).
5. J. Iwai, University of Washington Preprint, UWSEA 92-06.

Wilson Loop on a Light-Cone Cylinder

Stephen S. Pinsky
Department of Physics, The Ohio State University, Columbus, OH 43210

Abstract

QCD without matter and quantized on a light-cone spatial cylinder is considered. For the gauge group SU(N) the theory has $N-1$ quantum mechanical degrees of freedom, which describe the color fux that circulates around the the spatial cylinder. In $1+1$ dimensions this problem can be solved analytically. I use the solution for SU(2) to compute the Wilson loop phase on the surface of the cylinder and find that it is equal to $g^2 area/4$. This result is different from the well known result for flat space. I argue that for SU(N) the Wilson loop phase for a contour on a light-cone spatial cylinder is $g^2(area)(N-1)/4$. The underlying reason for this result is that only the $N-1$ dimensional Cartan subgroup of $SU(N)$ is dynamical in this problem.

1 Introduction

The expectation value of the Wilson loop is an important quantity in gauge theories. This phase is a gauge invariant quantity that provides information about the long range behavior of a theory, however it is beyond the reach of weak coupling perturbation theory calculations. On the other hand for non-perturbative approaches the Wilson loop is an important object to consider. While for QCD in 3+1 dimensions the calculation of the Wilson loop is quite difficult, in 1+1 dimension it is much more tractable and in some problems can be calculated exactly.

The problem of pure glue QCD in 1+1 dimensions in the gauge SU(N) with periodic boundary conditions can be solved solved exactly [1,2] since it has only $N-1$ degrees of freedom which are independent of space and the problem is therefore a quantum mechanical rather than a true field theory problem. Nevertheless the problem is very interesting from a number of points of view. First the degrees of freedom are simply color flux loops that circulate around the entire spatial cylinder and as such they rely on the fact that the problem is formulated on a cylindrical topology. This problem is particularly interesting to people studing light-cone field theory since it is the only known gauge theory where the Hamiltonian takes exactly the same functional form in both the light-cone and equal-time formulations.

We will briefly review the formulation and solution of this problem, here using the light-cone gauge and light-cone quantization in the gauge $SU(2)$. We will solve for the wavefunctions and the energy eigenvalue of the problem. We will solve the equations of motion for the vector potential and use the solution to calculate the path integral of the vector potential around a closed loop on the surface of the cylinder which makes up space time in this problem. We then calculate the Wilson loop by taking the vacuum expectation value of this loop calculation.

There is an exact general result for the Wilson loop expection value in 1+1 dimensional QCD in the absence of matter. The result we find here by direct calculation for $SU(2)$ on a cylinder does not agree with this result. This is perhaps not surprising because of the special topology of the space we consider. Based on our result for $SU(2)$ we sugest a general result for the value of the Wilson loop phase for $SU(N)$ on a cylinder. Our conjecture agrees with the general result in the large N limit as would be suggested by the work of Gross[5].

2 Gauge Fixing

The Lagrangian density for SU(2) non-Abelian gauge theory in 1+1 dimensions is,

$$\mathcal{L} = \frac{1}{2}\operatorname{Tr}\left(F_{\mu\nu}F^{\mu\nu}\right) \tag{1}$$

where $F_{\mu\nu} = \partial_\nu A_\nu - \partial_\nu A_\mu - g[A_\mu, A_\nu]$. We consider the theory on a finite interval, x^- from $-L$ to L, and we impose periodic boundary conditions on all gauge potentials A_μ.

We now show that the light-cone gauge $A^+ = 0$ which is the one that one normal would prefere to use for light-cone quantization cannot be reached. A gauge transformation U bringing a gauge potential B^μ, itself in some arbitrary gauge configuration, to some other gauge configuration A^μ is

$$gA^\mu = \partial_\mu U U^{-1} + gUB^\mu U^{-1} \,. \tag{2}$$

Here g is the coupling constant and U is an element of the Lie algebra of SU(2). Clearly U given by

$$U = P\exp\left[-g\int_{-L}^{x^-} dy^- B^+(y^-)\right] \tag{3}$$

will bring us to the gauge $A^+ = 0$.

We appear to have been successful in getting the light-cone gauge. However, the element U through which we wish to achieve the gauge condition must satisfy Z_2 periodic boundary condition, $U(x^-) = (\pm)U(x^- + 2L)$. This is so, because gauge fixing is usually done with trivial elements of the gauge group — namely transformations generated by the Gauss law operator via the classical brackets or corresponding quantum commutators. However for this to be carried through, one needs to be able to discard surface terms. With nonvanishing boundary conditions this can only be realized by Z_2 periodic elements U. Clearly Eq.(3) does not satisfy these boundary conditions. So in fact the attempt has failed.

With a modification of Eq.(3),

$$U(x) = e^{gx^- V^+} P e^{-g\int_{-L}^{x^-} dy^- B^+(y^-)} \,. \tag{4}$$

where V^+ is the the integral B^+ over space normalized by the length of the spatial cylinder, sometimes called the "zero mode", this is an allowed gauge transformation. However it does not completely bring us to the light-cone gauge. We find instead

$$A^+ = V^+ . \tag{5}$$

In other words, we cannot eliminate the zero mode of the gauge potential. The reason is evident: it is *invariant* under periodic gauge transformations. But of course we can always perform a rotation in color space. In line with other authors [10], we choose this so that $V^+ = v(x^+)\tau_3$ is the only non-zero element, since in our representation only τ^3 is diagonal.

In addition, we can impose the subsidiary gauge condition that the zero mode of A_3^- is zero. This would appear to have enabled complete fixing of the gauge. This is still not so. Gauge transformations

$$G = \exp\{ix^-(\frac{n\pi}{L})\tau^3\} \tag{6}$$

generate shifts, according to Eq.(2), in the zero mode component

$$v(x^+) \to v(x^+) + \frac{n\pi}{gL} . \tag{7}$$

All of these possibilities, labelled by the integer n, of course still satisfy $\partial_- A^+ = 0$, but as one sees $n = 0$ should not really be included. One notes that the transformation is x^--dependent and Z_2 *periodic*. It is thus a simple example of a Gribov copy [9] in 1+1 dimensions. We follow the conventional procedure by demanding

$$v(x^+) \neq \frac{n\pi}{gL} , \quad n = \pm 1, \pm 2, \ldots . \tag{8}$$

This eliminates singularity points at the Gribov 'horizons' which in turn correspond to a vanishing Faddeev-Popov determinant [8].

The equations of motion for the theory are

$$[D^\mu, F_{\mu\nu}] = \partial^\mu F_{\mu\nu} - g[A^\mu, F_{\mu\nu}] = 0 . \tag{9}$$

For our purposes it is convenient to break this equation up into color components A_a^μ. Color will always be the lower index. Rather than the three color fields A_1^μ, A_2^μ and A_3^μ we will use chiral notation with $A_+^\mu = A_1^\mu + iA_2^\mu$ and $A_-^\mu = A_1^\mu - iA_2^\mu$. With the above gauge conditions the $\nu = +$ equations are

$$(i\partial^+)^2 A_3^- = 0, \tag{10}$$
$$(i\partial^+ + gv(x^+))^2 A_-^- = 0 \tag{11}$$

These equation are of course easily solvable. The solution for A_3^- is zero up to a constant which is the zero mode. Earlier we used our gauge freedom to set this zero mode to zero. The operator in the equation for A_-^- is in fact not singluar in a particular Gribov region and is therefore invertible giving $A_-^- = 0$.

The only remaining equation of motion that is not totally trival is the equation for v

$$\partial_+^2 v(x^+) = 0 \tag{12}$$

The solution is of course

$$v(x^+) = \frac{g^2}{2\pi}\Pi_z x^+ + v(0) \tag{13}$$

and where Π_z is the canonical momentum defined below.

The Hamiltonian for this quantum mechanics problem is easily obtained from the above Lagrangian and we find,

$$P^- = L\partial_+^2 v(x^+) \tag{14}$$

This leads to a set of properly normalized conjugate variables,

$$z = \frac{gLv}{\pi} \quad \Pi_z = \frac{2\pi}{g}\partial_+ v \tag{15}$$

which satisfy thecanonical commution relation $[z, \Pi] = i$ in the fundamental modular domain $-1 < z < 0$. The Schrödinger equation is straightforward to solve and wavefunction and energy eigenvalues are

$$\psi_n(z) = C_n sin(n\pi z) \quad E_n = \frac{g^2 L(n^2-1)}{4} \tag{16}$$

where we have renormalized the ground state ($n = 1$) energy to zero and C_n are normalization constants. The wavefuntion must vanish at $z = 0$ and -1 the Gribov horizions[2].

3 Wilson Loop

The vacuum expection value of the Wilson loop provides information about the large distance behavior of a theory which is not accessable to pertubation theory calculation. The well established lore associated with the Wison loop is that if the phase goes like the area of the enclosed contour the theory is confining, whereas if the phase goes like the perimeter of enclosed contour the theory is not confining.

In 1+1 dimensions the general result for $SU(N)$ QCD without matter is [4],

$$W \propto e^{ig^2 \frac{(N^2-1)}{4N}\mathcal{A}}, \tag{17}$$

where $\mathcal{A}$ is the area enclosed by the Wilson loop. The Wilson loop for the problem we are considering here can be written as the vacuum expection value of the WIlson loop phase factor,

$$W = \langle\psi_1|TrPe^{ig\oint A\cdot dx}|\psi_1\rangle. \tag{18}$$

The vacuum expection value here takes the form of the expectation value with the ground state wavefunction. The contour that we will chose for the path integral consists of straight lines connecting the following points in (x^-, x^+) space on the surface of the light-cone space-time cylinder.

$$(0,0) \rightarrow (l,0) \rightarrow (l,t) \rightarrow (0,t) \rightarrow (0,0)$$

The only component of A^μ that is non-zero is $A^+ = v(x^+)\tau_3$; therefore the contour intergal yields $(v(0) - v(t))l\tau_3$. Now using the solution of the equation of motion for $v(x^+)$ we find for the contour integral,

$$ig \oint A \cdot dx = -ig^2 \; 2\pi \mathcal{A} \Pi_z \tau_3 \tag{19}$$

where $\mathcal{A}$ is the area of the enclosed contour. This leads to the following expression for the Wilson loop W,

$$W = Tr \int_{-1}^{0} dz sin(\pi z)(cos(\theta) + i\tau_3 sin(\theta))sin(\pi z) \tag{20}$$

where

$$\theta = i\frac{g^2 \mathcal{A}}{4\pi}\frac{d}{dz}$$

The momentum operator acts on the ground state wave function to the right. The expansion of the $sin(\theta)$ gives an odd number of derivatives leaving an integral of $sin(\pi z)cos(\pi z)$ which vanishes when integrated over the Gribov region $-1 < z < 0$. This leaves only the $cos(\theta)$ function and after some algebra I find

$$W \propto cos(\frac{g^2 \mathcal{A}}{4}) \tag{21}$$

The general result Eqn(17) for the Wilson loop when evaluated for $N = 2$ gives $\frac{g^2 3\mathcal{A}}{8}$.

4 Discussion

Let us first summarize the essential points. I analyzed pure glue non-Abelian gauge theory in a compact spatial volume with periodic boundary conditions on the gauge potentials. Working in the light-cone Hamiltonian approach, I demonstrated how one carefully fixes the gauge. The quantum field theory problem then reduces to a quantum mechanical problem which can be solved exactly. Given this exact non-perturbative result for the vacuum state and vector potential it becomes a straightforward calculation to evaluate the Wilson loop and the result for the gauge group $SU(2)$, $\frac{g^2 \mathcal{A}}{4}$ does not agree with the general result.

How can we understand these different results? The natural explanation seems to be that on the cylinder the gauge field only has support on the abelian Cartan subalgebra whereas the general result gets contributions from all color components. We can speculate about the extension of this calculation to SU(N) where the vector potential only has support only $N - 1$ dimensional abelian Cartan sub-algerbra. Since the contributions are abelian we expect the phases to simply add for each additional field component and therefore the Wilson loop phase should be

$$W \propto e^{\frac{ig^2 (N-1)\mathcal{A}}{4}}. \tag{22}$$

In a rather different context Gross [5] has identified the topological expansion of the space on which the Wilson loop is calculated with the $1/N$ expansion of the result. This then allows us to connect our calculation on a cylindrical topology with the large N expansion of the general result Eqn(17). We see that in the large N limit Eqn(17) and Eqn(22) agree. Thus a possible intepretation of this calculation might be the explicit verification of the Gross [5] result.

Acknowledgments

The author would like to acknowledge D. Robertson as a collaborator on most of this work. This work is supported by a grant from the US Department of Energy.

References

1. N. Manton, Ann. Phys. (NY) **159**, 220 (1985). J.E. Hetrick, Y. Hosotani, Phys. Rev. D**38**, 2621 (1988); J.E. Hetrick, Y. Hosotani, PLB **30** (1989) 88, F. Palumbo, Phys. Lett. B**243**, 109 (1990); S.G. Rajeev PLB **212** (1988) 203; E. Langmann, G.W. Semenoff, Phys. Lett. B**303**, 303 (1993).
2. A. C. Kalloniatis, H.C. Pauli and S.S. Pinsky Phys. Rev. **D50**, 6633 (1994).
3. J.E. Hetrick, Nucl. Phys. B (Proc. Suppl.) **30** 228 (1993), J.E. Hetrick UvA-ITFA 93-15 (hep-th/ 9305020).
4. A. Migdal, Sov. Phys. JETP **42** (1976) 413.
5. D. Gross, Nucl. Phys. **4 400** (1993) 161
6. S. J. Brodsky, G. McCartor, H. C. Pauli, S. S. Pinsky, Particle World **3**, 109 (1993).
7. M. Lüscher, Nucl. Phys. B**219**, 233 (1983); M. Lüscher, G. Münster, Nucl. Phys. B**232**, 445 (1984).
8. P. van Baal, Nucl. Phys. B**369**, 259 (1992) and references therein.
9. V.N. Gribov, Nucl. Phys. B**139**, 1 (1978); H. Yabuki, Phys. Lett. B**231**, 271 (1989).
10. V.A. Franke, Y.V. Novozhilov, E.V. Prokhvatilov, Lett. Math. Phys. **5**, 437 (1981); F. Lenz, H.W.L. Naus, M. Thies, 'QCD in the Axial Gauge Representation', Erlangen preprint. To appear in Ann. Phys.(NY) (1994).

Vortices and Confinement in Lattice Gauge Theory

Richard W. Haymaker
Department of Physics and Astronomy, Louisiana State University, Baton Rouge, Louisiana 70803, USA

We seek to exploit the connection between the spontaneous gauge symmetry breaking and vortices in superconductivity in order to uncover the mechanism for confinement in non Abelian lattice gauge theories.

1 Introduction

I wish to express my appreciation to Pete Carruthers for being my mentor when I was starting out. I was fortunate to be a postdoc with Pete at Cornell. He was pushing us to study spontaneous symmetry breaking in the early 70's, and further that we should look to condensed matter physics for inspiration. My talk has precisely these origins – the physics of superconductivity applied to color confinement[a].

The key principle governing the onset of superconductivity is the spontaneous breaking of the U(1) gauge symmetry via a non-zero vacuum expectation value of a charged field [1]. An immediate consequence of this is the generation of a photon mass and, for type II superconductors, the formation of vortices which confine magnetic flux to narrow tubes as revealed by the Ginzburg-Landau theory[2]. Lattice studies of dual superconductivity in SU(N) gauge theories seek to exploit this connection in establishing the underlying principle governing color confinement[3].

2 Abelian Theory

On a four dimensional lattice, the effective Euclidean lattice Higgs theory is the appropriate generalization of the Ginzburg-Landau theory. The Higgs field is a 0-form living on the sites and the gauge field is a 1-form living on the links. Classical solutions to this theory exhibit the connection between the non-zero vacuum expectation value of the Higgs field and vortex formation.

In U(1) lattice pure gauge theory (no Higgs field), this same connection is seen to be present, not in the defining variables, but rather in the dual variables. More specifically:

[a]Talk given at the workshop on "The state of physics at the end of the 20th century," in honor of Pete Carruthers' 61st Birthday.

1. A field with non-zero magnetic monopole charge, Φ, has been constructed[4]. It is a composite 4-form living on hypercubes constructed from gauge fields. There are also monopole current 3-forms.

   On the dual lattice this monopole operator is a 0-form living on dual sites. The monopole currents are 1-forms living on dual links. These currents either form closed loops or end at monopole operators.

   The monopole operator has a non-zero vacuum expectation value in the dual superconducting phase, $\langle\Phi\rangle \neq 0$, thereby signaling the spontaneous breaking of the U(1) gauge symmetry.

2. Dual Abrikosov vortices have been seen in simulations[5]. They are identified by the signature relationship between the electric field and the curl of the monopole current in the transverse profile of the vortex. The dual coherence length, ξ_d measures the characteristic distance from a dual-normal-superconducting boundary over which the dual-superconductivity turns on. The dual London penetration length, λ_d measures the attenuation length of an external field penetrating the dual-superconductor. The dual photon mass $\sim 1/\lambda_d$ and the dual Higgs mass $\sim 1/\xi_d$.

A signal $\langle\Phi\rangle \neq 0$ without the consequent signal of a dual photon mass does not imply confinement. An observation of a dual photon mass, i.e. vortex formation, without $\langle\Phi\rangle \neq 0$ does not reveal the underlying principle governing the phenomenon.

The lattice Higgs theory, treated as an effective theory, i.e. limited to classical solutions, and considered in the dual sense, provides a model for interpreting simulations of the pure gauge theory that can reveal these important connections[b]. On the lattice, the Higgs theory action is given by

$$\begin{aligned} S &= \beta \sum_{x,\mu>\nu} (1 - \cos\theta_{\mu\nu}(x)) \\ &\quad -\kappa \sum_{x,\mu} (\phi^*(x) e^{i\varphi_\mu(x)} \phi(x+\epsilon^{(\mu)}) + H.c.) + \sum_{x} V_{Higgs}(|\phi(x)|^2), \end{aligned} \tag{1}$$

where $\theta_{\mu\nu}(x)$ is the curl of the gauge field,

$$\theta_{\mu\nu}(x) = \Delta^+_\mu \varphi_\nu(x) - \Delta^+_\nu \varphi_\mu(x), \tag{2}$$

and where $\phi(x)$ is the Higgs field and $\phi(x+\epsilon^{(\mu)})$ refers to the Higgs field at the neighboring site in the μ direction. The electric current is given by

$$\frac{a^3}{e\kappa} J^e_\mu(x) = Im(\phi^*(x) e^{i\varphi_\mu(x)} \phi(x+\epsilon^{(\mu)})), \tag{3}$$

[b]This is described in Ref. 2 which we review briefly here.

where a is the lattice spacing. Let us choose a Higgs potential that constrains the Higgs field $|\phi(x)| = 1$. Then if

$$\sin[\theta + 2N\pi] \approx \theta, \tag{4}$$

we find a relation between the field strength tensor and the curl of the current

$$\mathcal{F}_{\mu\nu} \equiv F_{\mu\nu} - \frac{a^2}{e^2\kappa}\frac{(\Delta_\mu^+ J_\nu^e(x) - \Delta_\nu^+ J_\mu^e(x))}{a} = \frac{2\pi N}{e}\frac{1}{a^2} = Ne_m\frac{1}{a^2}, \tag{5}$$

where

$$ea^2 F_{\mu\nu} = \sin[\varphi_\mu(x) + \varphi_\nu(x + \epsilon^{(\mu)}) - \varphi_\mu(x + \epsilon^{(\nu)}) - \varphi_\nu(x)]. \tag{6}$$

If $N = 0$ then this is a London relation which implies a Meissner effect. For $N \neq 0$ there are N units of quantized flux penetrating that plaquette, indicating the presence of an Abrikosov vortex.

The key point is that the modulus of the Higgs field must be independent of position to get this relation, leaving the phase of the Higgs field as the relevant dynamical variable, i.e. spontaneous gauge symmetry breaking. For a "Mexican hat" Higgs potential, there is a coherence length setting the scale over which the vacuum expectation value of the Higgs field reaches its asmptotic value.

In simulations of a pure U(1) gauge theory we see the same relation but in the dual variables[c].

$$^*\mathcal{F}_{\mu\nu} \equiv {}^*F_{\mu\nu} - \lambda_d^2\frac{(\Delta_\mu^- J_\nu^m(x) - \Delta_\nu^- J_\mu^m(x))}{a} = Ne\frac{1}{a^2}. \tag{7}$$

3 Non-Abelian Theory

The link of these considerations to confinement in non-Abelian gauge theory is through the Abelian projection[6]. One first fixes the gauge while preserving U(1) gauge invariance. The non-Abelian gauge fields can be parametrized in terms of a U(1) gauge field and charged coset fields. The working hypothesis is that operators constructed from the U(1) gauge field alone, i.e. Abelian plaquettes, Abelian Wilson loops, Abelian Polyakov lines and monopole currents, will exhibit the correct large distance correlations relevant for confinement. But there is as yet no definitive way to choose the gauge which defines the Abelian projection and hence no unique way to define Abelian links and coset fields from the SU(2) links.

---

[c]See Ref 2 for a review of these results

Baker, Ball, Zachariasen[7] and co-workers have postulated a dual form of QCD in the continuum. Their formulation describes a non-Abelian dual superconductor and hence it confines color. They have calculated flux tubes, static quark potentials, temperature dependent effects and many other quantities in the tree approximation. Although their formulation does not focus on monopoles, there is a large potential overlap between their work and the lattice description developed here.

We are seeking a judicious choice of dynamical variables — defined by a particular Abelian projection — which can account for confinement via an effective theory involving these dynamical variables. This should be our first goal. If this is solved satisfactorily, then we can investigate how the picture changes if we go to an alternative set of variables — a different Abelian projection — since the phenomenon we are describing is of course gauge invariant. We do not expect that the same mechanism would describe confinement in two different Abelian projections.

There is a very nice illustration of this point in a paper by Chernodub, Polikarpov and Veselov[8]. They compare two Abelian projections.

- The first is the maximal Abelian gauge[9] which is the most widely studied and perhaps the most promising candidate. Monopoles are the magnetic charge carriers of the persistent currents.

- Second they exhibit an Abelian projection in which confinement is due to objects other than monopoles. They choose the "minimal Abelian projection" and show that confinement is due to topological objects which are denoted "minopoles".

Here we have two projections, two sets of dynamical variables, and two different descriptions of confinement, both viable candidate mechanisms.

There is an intimate connection between vortex formation and a non-zero vacuum expectation of the monopole field in a Higgs theory and hence it is a strong test of the idea for the dual theory. The two obvious candidate projections are the Polyakov gauge and the Maximal Abelian gauge.

Polyakov Gauge: Del Debbio, Di Giacomo, Paffuti and Pieri[10] have constructed a lattice monopole field operator that shows a very strong signal with a sharp discontinuity in the vacuum expectation value of the monopole operator as a function of temperature in the neighborhood of T_c, the deconfining temperature. Their calculation is manifestly gauge invariant.

The monopole field operator creates the Abelian magnetic charge and the accompanying Abelian field strength. The definition of gauge invariant field strength is that defined by 'tHooft[11] in the Georgi-Glashow model,

$$\begin{aligned} F_{\mu\nu} &= Tr\Big[\phi\Big(\partial_\mu A_\nu - \partial_\nu A_\mu - ig[A_\mu, A_\nu]\Big)\Big] \\ &+ \frac{1}{2}\ Tr\Big[\phi\Big(\partial_\mu\phi - ig[A_\nu,\phi]\Big)\Big(\partial_\nu\phi - ig[A_\mu,\phi]\Big)\Big], \end{aligned} \quad (8)$$

where ϕ transforms under the adjoint representation of SU(2). However there is no adjoint scalar field in QCD. Instead one constructs a composite field from a Wilson line, which begins and ends at the same site x

$$W(x) = \cos\theta(x) + i\phi(x)\sin\theta(x), \quad \phi = \widehat{\phi}\cdot\vec{\sigma}, \quad \widehat{\phi}\cdot\widehat{\phi} = 1, \quad \theta \in [0,\pi]. \quad (9)$$

Del Debbio et. al. chose $W(x) = P(x)$, a Polyakov line which closes through the time dimension. The Polyakov gauge is obtained by rotating $\widehat{\phi}(x)$ into the the 3−direction at every site. Then the gauge invariant field strength reduces to curl of the the 3rd component SU(2) field tensor. Even though the construction is gauge invariant we associate their calculation with the Polyakov gauge. In Bernstein et.al.[12] we established vortex formation in the Polyakov gauge.

Maximal Abelian gauge: The maximal Abelian gauge is defined by a gauge configuration that maximizes $\mathcal{R}$, where

$$\mathcal{R}[U] \equiv \sum_{x,\mu}\frac{1}{2}Tr\left(\sigma_3 U_\mu(x)\sigma_3 U_\mu^\dagger(x)\right), \quad (10)$$

and where $U_\mu(x)$ is the link starting a site x and extending in the μ direction. In the continuum limit this becomes the gauge condition

$$\left(\partial_\mu \pm gA_\mu^3(x)\right) A_\mu^\pm(x) = 0. \quad (11)$$

The vortices are well established[13,12] in this gauge. The typical behavior is shown in Fig. 1. The London relation is seen in the confining case for transverse distances larger than about two lattice spacing. The dual coherence length $\xi_d \approx 2$, i.e. the onset of the violation of the London relation. The unconfined case is also shown where curl J is almost zero, and the electric field falls more gradually than in the confining case.

The static potential constructed from Abelian links gives as definitive a signal of confinement as the gauge invariant static potential as found by Suzuki et.al.[3,14], Stack et.al.[15] and Bali et. al.[16]. Bali et.al. find

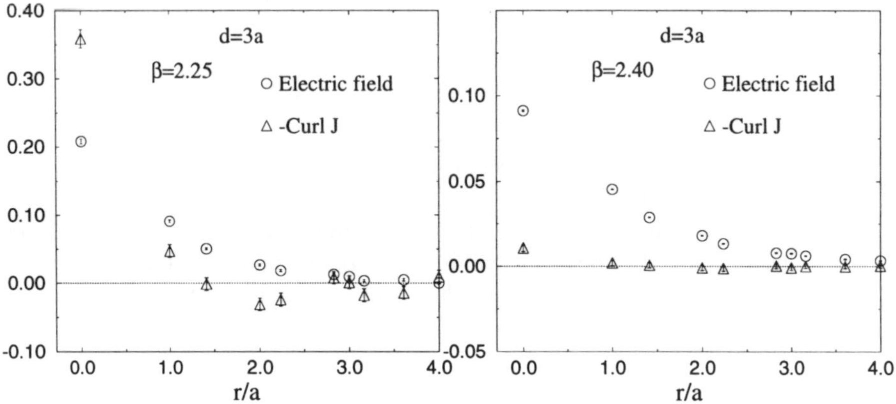

Figure 1: Transverse profile of the electric field and curl of the monopole current in the mid plane between a static $q\bar{q}$ pair ion the maximal Abelian gauge at finite temperature for a confining (left) and unconfining (right) phases. Taken from Ref. 8

the Abelian string tension calculation gives 0.92(4) times the full string tension for $\beta = 2.5115$. Whether this approaches 1.0 in the continuum limit remains to be seen.

The calculation of Del Debbio et.al[10] of the monopole field operator as a practical matter is not adaptable to this gauge. It would require hundreds of gauge fixing sweeps of the whole lattice in order to accept or reject a single link update. Chernodub, Polikarpov and Veselov[17] have more recently calculated the constraint effective potential as a function of the monopole field in this gauge and found a symmetry breaking minimum. However they reported a problem of obtaining statistics and instead calculated an approximation to the effective potential.

More recently Nakamura et al. [18] studied an alternative monopole operator defined in terms of the variables which occur in the monopole form of the action. We have not yet addressed the issue of establishing vortex formation in that framework.

4 Conclusions

A linearly rising static quark potential is ample evidence that quarks are confined. However this is a global calculation, not revealing whether there is a simple underlying mechanism governing the physics of confinement. Calculations of the distribution of energy density surrounding a quark-antiquark pair have shown that there is a flux tube between quarks as the linear potential implies[19,20]. This is a calculation of local properties which reveals detail of the confinement configurations. The present effort is to find local relationships among field strengths and currents pointing toward the existence of an effective theory of confinement. The maximal Abelian projection remains the most promising approach to explain quark confinement in these lattice studies. However it may run into difficulty in explaining color confinement more generally, for example in the case of static sources transforming under the adjoint representation[21]. Nevertheless the lattice studies have opened up techniques to investigate the seminal ideas that were proposed long ago by Nielsen and Olesen[22], Kogut and Susskind[23], Parisi[24], Mandelstam[25] and 'tHooft[26].

5 Acknowledgments

Work was partially supported by the Department of Energy under Grant No. DE-FG05-91ER40617

References

1. S. Weinberg, The Quantum Theory of Fields Vol II,Cambridge University Press, Cambridge 1996, See chapter 21.6.
2. For a review and further references to superconductivity on the lattice see e.g. R.W. Haymaker, *"Dual Abrikosov vortices in U(1) and SU(2) lattice gauge theories"*, Proceedings of the international School of Physics "Enrico Fermi", Course CXXX, A. Di Giacomo and D. Diakonov (Eds.) IOS Press, Amsterdam 1996, hep-lat 9510035.
3. For a review of monopoles and confinement, see e.g. T. Suzuki, Nucl.Phys. **B** Proc. Suppl. **30**, 176 (1993).
4. J. Frölich and P. A. Marchetti, Euro. Phys. Lett. **2**, 933 (1986); Commun. Math. Phys. **112**, 343 (1987); **116**, 127 (1988); **121**, 177 (1989); Lett. Math. Phys **16**, 347 (1988); L. Polley and U.-J. Wiese, Nucl. Phys. **B356**, 621 (1991); M. I. Polikarpov, L. Polley and U.-J. Weise, Phys. Lett. **B253**, 212 (1991);. L. Del Debbio, A. Di Giacomo and G. Paffuti, Phys. Lett. **B349**, 513, (1995); Nucl. Phys B (Proc. Suppl.) **42**, 231 (1995).
5. V. Singh, R.W. Haymaker, and D.A. Browne, Phys. Rev. **D47**, 1715 (1993).
6. G. 't Hooft, Nucl. Phys. **B 190**, 455 (1981).

7. M. Baker, J. S. Ball and F. Zachariasen, Physics Reports, **209** No.3, 73 (1991); Phys. Rev. **51**, 1968 (1995);
8. M.N. Chernodub, M.I. Polikarpov and A.I. Veselov, Phys. Lett. **B 342**, 303 (1995).
9. A.S. Kronfeld, G. Schierholz and U.J. Wiese, Nucl. Phys. **B 293**, 461 (1987); A.S. Kronfeld, M.L. Laursen, G. Schierholz and U.J. Wiese, Phys. Lett. **B 198**, 516 (1987); T. Suzuki and I. Yotsuyanagi, Phys. Rev. **D 42**, 4257 (1990).
10. L. Del Debbio, A. Di Giacomo, G. Paffuti and P. Pieri, Phys. Lett. **B 355**, 255 (1995).
11. G. 't Hooft, Nucl. Phys. **B 79**, 276 (1974).
12. K. Bernstein, G. DiCecio, and R. W. Haymaker, LSU Preprint LSUHEP No. 213-1996.
13. V. Singh, D.A. Browne, and R.W. Haymaker, Nucl. Phys. B (Proc. Suppl.) **30**, 658 (1993); Phys. Lett. **B306**, 115 (1993); Y. Matsubara, S. Ejiri, and T. Suzuki, Nucl. Phys. **B34** (Proc. Suppl.) 176 (19940; Y. Peng and R.W. Haymaker, Phys. Rev. **D 52** 3030, (1995).
14. H. Shiba and T. Suzuki, Phys. Lett. **B 333**, 461 (1994).
15. J. Stack and R. Wensley, Nucl. Phys. **D 22**, 597 (1992); J. Stack, S. Neiman and R. Wensley, Phys. Rev. **D 50**, 3399 (1994).
16. G. S. Bali, V. Bornyakov, M. Müller-Preussker and K. Schilling, Phys. Rev. **D 54** 2863 (1996).
17. M.N. Chernodub, M.I. Polikarpov and A.I. Veselov, Nucl. Phys. Proc. Suppl. **49**, 307, (1996).
18. N. Nakamura, V. Bornyakov, S. Ejiri, S. Kitahara, Y. Matsubara and T. Suzuki, Nucl. Phys.B (Proc. Suppl.)**53**, 512 (1997).
19. R. W. Haymaker, V. Singh, Y. Peng, and J. Wosiek, Phys. Rev. **D 53**, 389 (1996).
20. G. S. Bali, C. Schlichter, and K. Schilling, Phys. Rev. bf D 51, 5165 (1995).
21. L. Del Debbio, M. Faber, J. Greensite, and S. Olejnik, Preprint LBNL-39424 hep-lat/9610005; Nucl. Phys. **B 53**, 141 (1997).
22. H.B. Nielsen and P. Olesen, Nucl. Phys. **B61**, 45 (1973).
23. J. Kogut and L. Susskind, Phys. Rev. D **9**, 3501 (1974).
24. G. Parisi, Phys. Rev. **D11**, 970 (1975),
25. S. Mandelstam, Phys. Rep. **23** 245,(1976). 5
26. G. 'tHooft, High energy physics, ed. A. Zichichi (Editrice Compositori, Bologna, 1976).

Nonuniformity of the $1/N$ Expansion for Two-Dimensional $O(N)$ Models

Adrian Patrascioiu* **and Erhard Seiler**

Max-Planck-Institut für Physik, Werner-Heisenberg-Institut
Föhringer Ring 6, 80805 Munich, Germany

Abstract

We point out that the $1/N$ expansion, which is widely invoked to infer properties of the $2D$ $O(N)$ models, is nonuniform in the temperature, i.e. with decreasing temperature the $1/N$ expansion truncated at a fixed order deviates more and more from the true answer. This fact precludes the use of the expansion to deduce low temperature properties such as asymptotic scaling for those models. By contrast, in the $1D$ $O(N)$ chains, there are no signs of such a nonuniformity.

* Permanent address: Physics Department, University of Arizona, Tucson AZ 85721, U.S.A.

1 Introduction

The $1/N$ expansion [1] is a very popular tool to analyze $2D$ $O(N)$ models, especially in the critical dimension 2. But already the first rigorous paper by A.Kupiainen [2], which established the fact that the expansion is a valid asymptotic expansion in the thermodynamic limit at fixed $\tilde{\beta} = \beta/N$, contained hints that the expansion is really only good at high enough temperature: the main result stated that the expansion is valid provided $\tilde{\beta}$ is less than the critical value of the 'spherical model', i.e. the $N \to \infty$ limit of the model. Likewise the followup paper by Fröhlich et al [4] proved Borel summability of the expansion only for small $\tilde{\beta}$. Müller, Raddatz and Rühl [5] computed the $1/N$ corrections to the mass gap and the susceptibility for large $\tilde{\beta}$; their results are nonuniform in $\tilde{\beta}$ in the sense that the corrections grow beyond all bounds as $\tilde{\beta} \to \infty$. This fact, which is apparent from the equations in [5] was, however, not noted explicitely by the authors. Finally, U.Wolff computed the correlation length of the $O(N)$ model for $N = 2, 3, 4, 8$. Comparing his results with the spherical model makes it quite obvious that the approach to the $N \to \infty$ limit slows down with increasing $\tilde{\beta}$ (see Tab.1).

In spite of all those facts, many authors (e.g.[1, 6, 7]) continue to cite the $1/N$ expansion and the properties of the spherical model as evidence for the supposed asymptotic scaling and absence of a critical point of the $2D$ $O(N)$ model at finite N. In this note we report a few more calculations that show explicitly that the $O(1/N)$ corrections for a number of quantities grow linearly with $\tilde{\beta}$. All those quantities are of long range character, such as correlation lengths, susceptibilities or two-point functions at large distances. We did not see such effects in short range quantities such as the energy (= two point function at one lattice distance); this does not mean, however, that the $1/N$ expansion for short range quantities is uniform in $\tilde{\beta}$. Uniformity would require control of the deviation of the true result from its truncated $1/N$ expansion, something that is not analyzed here. We should, however, point out that our recent results on the problems of perturbation theory (PT) [8] in non-Abelian models suggest that in fact there is nonuniformity even in those short-distance quantitities.

The situation is different in $1D$: no indication of any non-uniformity is found. The underlying reason is that in $1D$ the necessary computations can be reduced to the analysis of simple 1 dimensional integrals, so that no infrared effects, which in our mind are responsible for the nonuniformity in $2D$, can arise.

2 Notation and Definitions

Consider the lattice $\mathbb{Z}^D$ and let Λ be some given finite subset of it. The non-linear $O(N)$ σ-model with standard nearest neighbor interaction (s.n.n.i.) at inverse temperature $\beta = \tilde{\beta}N$ is defined by the generating function

$$Z_\Lambda(g) = \int \prod_{\langle ij\rangle} \exp(\beta s_i \cdot s_j) \prod_i \Big(\exp(g_i \cdot s_i)\delta(s_i^2 - 1)ds_i \Big) \tag{1}$$

or equivalently the Gibbs measure

$$\frac{1}{Z_\Lambda(0)} \prod_{\langle ij\rangle} \exp(\beta s_i \cdot s_j) \prod_i \Big(\delta(s_i^2 - 1)ds_i \Big) \tag{2}$$

Here s is a N-vector, the integration is over the spins inside Λ with some boundary conditions.

It is well known (see for instance [2]) how one can develop an asymptotic expansion in powers of $1/N$ by introducing a 'Langrange multiplier field' a dual to the δ-function constraint and a standard saddle point expansion. In terms of the Lagrange multiplier field one obtains a dual representation of the model; the generating function is now given by

$$Z(g) = \int \prod_i da_i \exp(-N\hat{S}) \exp\Big(\frac{1}{2\tilde{\beta}N} (g, (-\Delta + m^2 - \frac{2ia}{\sqrt{N}})^{-1} g) \Big) \tag{3}$$

where the dual action is given by

$$\hat{S} = \frac{1}{2} \mathrm{tr} \ln(-\Delta + m^2 - \frac{2ia}{\sqrt{N}}) - i\tilde{\beta} \sum_i \frac{a_i}{\sqrt{N}} \tag{4}$$

The mass parameter m appearing in this equation is fixed by the gap equation

$$\tilde{\beta} = (-\Delta + m^2)^{-1}_{ii} \tag{5}$$

Kupiainen [2] showed, using the technique of cluster expansions, that the resulting asymptotic expansion is uniform in the volume Λ and can be used to obtain a valid asymptotic $1/N$ expansion in the infinite volume, provided $\tilde{\beta} < \tilde{\beta}_{sph}$, where $\tilde{\beta}_{sph}$ is the critical point of the spherical model. It is well known how the $1/N$ expansion can be cast in the form of Feynman graphs involving two types of propagators B and C given by

$$C_{ij} = (-\Delta + m^2)^{-1}_{ij} \tag{6}$$

$$B^{-1}_{ij} = \frac{1}{2}(C^2_{ij}) \tag{7}$$

Those Feynman graphs can be easily evaluated numerically on finite lattices. In this paper we are considering the $1/N$ corrections to the correlation length ξ, the susceptibility χ and the following ratio of two-point functions

$$A \equiv \frac{G(\xi_\infty/2)}{G(\xi_\infty/4)} \tag{8}$$

where ξ_∞ is the correlation length of the spherical model (up to a correction that is negligible for $\tilde{\beta} > .5$, $\xi_\infty = 1/m$ where m is the solution of the gap equation) and

$$G(x) \equiv \langle s(0) \cdot s(x) \rangle \tag{9}$$

The quantities ξ, χ, A as well as the energy E are expanded as follows:

$$\xi = \xi_\infty(1 + \frac{1}{N}\xi_1 + O(\frac{1}{N^2})) \tag{10}$$

$$\chi = \chi_\infty(1 + \frac{1}{N}\chi_1 + O(\frac{1}{N^2})); \tag{11}$$

$$\frac{G(\frac{1}{2}\xi_\infty)}{G(\frac{1}{4}\xi_\infty)} = A_\infty(1 + \frac{1}{N}a_1 + O(\frac{1}{N^2})) \tag{12}$$

$$E \equiv \langle s(0) \cdot s(1) \rangle = E_\infty(1 + \frac{1}{N}e_1 + O(\frac{1}{N^2}) \tag{13}$$

In the following we will study the behavior of the coefficients ξ_1, χ_1 and a_1 for large $\tilde{\beta}$.

3 What is the problem?

Let us consider a certain physical quantity $A(N, \tilde{\beta})$. According to the preceding discussion it will have an asymptotic expansion of the form

$$A(N, \tilde{\beta}) \sim \sum_{k=0}^{\infty} \frac{a_k(\tilde{\beta})}{N^k} \tag{14}$$

The meaning of the symbol $\sim$ is that the series on the right hand side, if truncated at the term of order N^{-k}, will approximate the left hand side by up to a remainder term which is $o(N^-k)$, and this is true for any $k \geq 0$. That means

$$r_k(N, \tilde{\beta}) \equiv A(N, \tilde{\beta}) - \sum_{k=0}^{k} \frac{a_k(\tilde{\beta})}{N^k} \tag{15}$$

satisfies

$$\lim_{N\to\infty} |r_k(N, \tilde{\beta})| N^k = 0 \tag{16}$$

Uniform asymtoticity means the stronger property

$$|r_k(N, \tilde{\beta})| N^k \leq c_k(N) \tag{17}$$

where $c_k(N)$ is independent of $\tilde{\beta}$ and satisfies

$$\lim_{N\to\infty} c_k = 0 \tag{18}$$

It is easy to see that uniform asymtopticity does not hold if any one of the expansion coefficients $a_k(\tilde{\beta})$ is unbounded for large $\tilde{\beta}$, but it should be stressed that this is not a necessary condition. There are simple examples of functions where every one of the coefficients $c_k(\tilde{\beta})$ is bounded and yet the expansion is not uniformly asymptotic. Consider for instance the function

$$f(N\tilde{\beta}) = \tilde{\beta}\exp(-N/\tilde{\beta}) \tag{19}$$

Its asymptotic expansion in powers of $1/N$ has all coefficients $a_k(\tilde{\beta}) = 0$ and the remainder term is always equal to f, which is unbounded.

4 One Dimension

In $1D$ the situation is particularly simple because the thermodynamic limit of the expectation values of local quantities can be obtained by computing in a finite box with free boundary conditions. For nearest neighbor quantities like the energy E one only has to evaluate a $1D$ integral; long range quantities such as the correlation length ξ or the susceptibility χ can be expressed in terms of E. It turns out that for this reason no nonuniformity in any of those quantities is found.

The energy of a $1D$ $O(N)$ chain can be obtained as follows:

$$E \equiv \langle s(0) \cdot s(1) \rangle = \frac{d}{d\beta} \ln Z \tag{20}$$

where Z is the partition function for one link:

$$Z = \int_{-1}^{1} dt e^{\beta t}(1-t^2)^{\frac{N-3}{2}} \tag{21}$$

Instead of using the general scheme outlined in the previous section, it is convenient to obtain the $1/N$ expansion directly from this expression. The two point function in the thermodynamic limit is simply given by

$$\langle s(0) \cdot s(n) \rangle = E^n \tag{22}$$

and hence the correlation length is

$$\xi = -\ln E \tag{23}$$

and the susceptibility

$$\chi = \frac{1+E}{1-E} \tag{24}$$

Introducing the rescaled inverse temperature

$$\bar{\beta} = \frac{\beta}{N-3} \tag{25}$$

the partition function takes the form

$$Z = \int_{-1}^{1} dte^{(N-3)f(t)} \tag{26}$$

with

$$f(t) = \bar{\beta}t + \frac{1}{2}\ln(1-t^2) \tag{27}$$

It is now straightforward to produce a saddle point expansion in powers of $1/(N-3)$: the saddle point value t_o is given by

$$t_o(\bar{\beta}) = \sqrt{1 + \frac{1}{4\bar{\beta}^2}} - \frac{1}{2\bar{\beta}} \tag{28}$$

and the logarithm of the partition function is expanded as

$$\ln Z = (N-3)f_{\bar{\beta}(t_o)} + \frac{1}{2}\ln\frac{2\pi}{(N-3)(-f''_{\bar{\beta}(t_o)})} + O(\frac{1}{(N-3)^2}) \tag{29}$$

Since t_o is a function that increases monotonically from 0 to 1 as $\bar{\beta}$ increases from 0 to ∞, we may use t_o instead of $\bar{\beta}$ as an independent variable. Using

$$\frac{d}{d\bar{\beta}} = \frac{(1-t_o^2)^2}{1+t_o^2}\frac{d}{dt_o} \tag{30}$$

the expansion for the energy becomes

$$E = t_o - \frac{t_o}{N-3}\frac{(1-t_o^2)(3+2t_o^2)}{1+t_o^2} + O(\frac{1}{(N-3)^2}) \tag{31}$$

It is already visible that the ration of the $1/(N-3)$ correction to the leading term remains bounded as $\bar{\beta} \to \infty$, i.e. $t_o \to 1$; in fact it is $O(1/\bar{\beta})$. But we want to expand in powers of $1/N$, not $1/(N-3)$, hence we have to re-expand the leading term, using

$$\tilde{\beta} = \frac{\bar{\beta}}{1-3/N} \tag{32}$$

Again it is very easy to see that we just obtain another correction of order $1/\tilde{\beta}$ to the $1/N$ contribution e_1 (see eq.(13)). It is also easy to see that the result is in agreement with the one found by Hasenfratz [3] for the infinite chain.

Using this information and the formulae above it is now trivial to conclude that

$$\xi_1 = \frac{e_1}{\ln E} \tag{33}$$

$$\chi_1 = -\frac{2e_1 E^2}{(1+E)(1-E)} \tag{34}$$

$$a_1 = -\frac{1}{2} e_1 \ln E \tag{35}$$

are all $O(1)$ for $\tilde{\beta} \to \infty$ as claimed.

5 Two Dimensions

We computed the Feynman graphs for the $1/N$ correction to the two point function numerically on lattices satisfying $L/\xi_\infty = 7$. It is observed that this is sufficient to reach the thermodynamic limit with a precision of 1 to 2%.

The expansion for the two point function has the form

$$G(i) = G_\infty(i) + \frac{1}{N} \sum_{j,l \in \Lambda} \frac{1}{L^4} G_\infty(i-j) \Sigma(j-l) G_\infty(l) \tag{36}$$

where $G_\infty(i) = C(i)/\tilde{\beta}$ and the 'self-energy' Σ is given by

$$\Sigma(i) = -B(i)G_\infty(i) + \frac{1}{2} \sum_{j,l,k \in \Lambda} \frac{1}{L^6} B(i-j)B(l-k)G_\infty(l-k)G_\infty(l-j)G_\infty(k-j) \tag{37}$$

and analogously in momentum space

$$\hat{G}(p) - \hat{G}_\infty(p)^2 \hat{\Sigma}(p) \tag{38}$$

where $\hat{\Sigma}$ is given by

$$\hat{\Sigma}(p) = -\sum_k \frac{1}{L^2} \hat{G}_\infty(k) \hat{B}(p-k) + \frac{1}{2} \hat{B}(0) \sum_{k,q} \frac{1}{L^4} \hat{G}_\infty(k)^2 \hat{G}_\infty(k+q) \hat{B}(q) \tag{39}$$

the momenta p, k, q range over $2\pi i/L$, $i = 1, 2, \ldots L$. ¿From the momentum space two-point functions $\hat{G}(0)$ and $\hat{G}(2\pi/L)$ we obtain the susceptibility and also the (effective) correlation length using the formula

$$\xi = \frac{1}{2 \sin \frac{\pi}{L}} \sqrt{\frac{\hat{G}(0)}{\hat{G}(\frac{2\pi}{L})} - 1} \tag{40}$$

which is expanded in powers of $1/N$ using the expansion for the two-point functions.

The results are given in Table 2. They show clearly that the $1/N$ corrections for all three quantities grow linearly with $\tilde{\beta}$, respectively $\ln m$, in agreement with the findings of [5].

6 Discussion

We have found by explicit calculation that the $1/N$ expansion is not uniformly asymptotic in $\tilde{\beta}$ for a variety of long range quantities in the $2D$ $O(N)$ model. The nonuniformity appears in the most blatant form: the correction grows linearly with $\tilde{\beta}$. For short distance quantities like the energy we could not find this kind of phenomenon up to $O(1/N^3)$. This does not mean, however, that the $1/N$ expansion in uniform for those quantities; to show uniformity one would have to control the remainder terms.

This situation is in striking contrast with the one for the $1D$ chains: since all compututations reduce to one-dimensional integrals, there is no nonuniformity occurring.

Some authors (for instance [6]) try to improve the quality of the $1/N$ expansion by introducing a certain 'mass renormalization'. Since we are working on a lattice, there is first of all no need to do so since there are no divergencies to be absorbed. Furthermore, the procedure corresponds to a a resummation of certain terms in the $1/N$ expansion and therefore is no longer a systematic asymptotic expansion in any parameter. While apparently improving the agreement with the Monte Carlo data in the $\tilde{\beta}$ region studied, the mathematical status of the procedure remains unclear and it can be expected to reveal the same problems as the systematic expansion at larger $\tilde{\beta}$.

In [8] we found that PT for the energy yields results that depend on the boundary conditions to order $1/\tilde{\beta}^2$; this dependence survives the limit $N \to \infty$. In our mind this is a strong indication that the $1/N$ expansion is nonuniform in $\tilde{\beta}$ even for short distance quantities, since the spherical limit of the model does not show this phenomenon.

But the main conclusion is that that the nonuniformity for the mass gap and similar quantities implies that it also holds for the Callan-Symanzik β-function. This means that the $1/N$ expansion cannot be used to draw any conclusions about the phase structure of the model for finite N, in particular the fact that the spherical model does not have a soft phase in $2D$ is in no way in conflict with the arguments advanced by us [9, 10] that strongly suggest the existence of such a phase for any finite N.

Note added in proof: After this paper was circulated as an e-print on the hep-lat bulletin board, a comment on it by M.Campostrini and P.Rossi appeared on the same bulletin board (hep-lat 9407008); this comment ques-

Table 1: *Comparison of Monte Carlo results with the spherical limit*; Monte Carlo data taken from [11]($O(2)$), [12] ($O(3)$), [13] ($O(4)$) and[14] ($O(4)$ and $O(8)$). The numbers represent ξ/ξ_∞.

$\tilde\beta$	.500	.5667	.575	.600	.650	.725
$N=8$	1.307		1.494		1.711	1.983
$N=4$	1.866		2.857	3.301	4.425	
$N=3$	2.644	5.479	8.348	15.629		
$N=2$	9.691	∞	∞	∞	∞	∞

Table 2: *The* $1/N$ *corrections to various long range quantities.* For the definition of the quantities given see eq. (10),(11),(12).

$\tilde\beta$	.49349	.60600	.67105	.71704	.78173	.82758
L	28	56	84	112	168	224
ξ_∞	4	8	12	16	24	32
ξ_1	1.764	2.927	3.640	4.152	4.879	5.397
χ_1	2.860	4.932	6.229	7.171	8.518	9.487
a_1	$-.486$	$-.853$	-1.058	-1.205	-1.416	-1.567

tioned the relevance of our observations, but in our view reflects again some widespread misconceptions about the issue. Therefore we would like to add the following remarks, which were also circulated as an e-print (hep-lat 9407030):

In our paper we tried to emphasize two points regarding the $1/N$ expansion:

1) Mathematically rigorous results obtained so far can establish *only* that it produces the correct asymptotic expansion at fixed $\tilde{\beta}$; in particular they do not guarantee that the expansion is asymptotic to the true expectation values uniformly in $\tilde{\beta}$ so that information about the true behavior for large $\tilde{\beta}$ can be obtained from it.

2) In $2D$, but not in $1D$, the coefficient of the $1/N$ term in the expansion of $G(\xi_\infty/2)/G(\xi_\infty/4)$ grows linearly with $\tilde{\beta}$. The quantity we consider is clearly 'adimensional', as they call it, but the fact that the coefficient of $1/N$ is unbounded implies that the expansion is non-uniform, contrary to their statement No.3.

Campostrini and Rossi have nothing to say about point (1), whose significance they seem to have missed completely. Indeed in the first part of their Comment, they discuss their results for the *continuum* $O(N)$ model. At the present time, no such model has been constructed. Consequently how could one ask whether the $1/N$ expansion produces the correct asymptotic expansion or not (see Sect.3 in our paper for the correct definition of asymptoticity), when the quantities with which the expansion is to be compared have not even been defined?

They acknowledge that nonuniformity does occur and call it a 'predicted property'. Presumably they mean that the asymptotic scaling formula itself has this property. But this fact certainly does not make the nonuniformity any less real. Similarly, they say that they can compute those nonuniformities analytically; again, that does not detract from the fact that they are there.

Finally, they claim that the nonuniformities in certain quantities disappear if one expands in $1/(N-2)$ instead of $1/N$. It is easy to see that this is mathematically impossible and that one can prove the following

Theorem: *Let $F(N,\beta)$ be a function that has an aymptotic expansion in powers of $1/(N-2)$ that is uniformly asymptotic in β. Then F has also an asymptotic expansion in powers of $1/N$ that is uniformly asymptotic.*

Campostrini and Rossi probably got mislead by the fact that by expanding in $1/(N-2)$ they can make certain manifestly nonuniform terms vanish. But uniform asymptoticity is a not a statement about some terms in the expansion, but about the truncation error.

The failure of Campostrini and Rossi to understand the issue is further illustrated by their statement: 'Adimensional ratios of physical quantities, computed in the lattice model, show uniform asymptoticity (actually complete independence of β in the scaling limit'. Again, as we stressed in our paper, to establish the correctness of the expansion one must bound the truncation error; the finiteness of the expansion coefficients for $\tilde{\beta} \to \infty$ is necessary, but not sufficient to establish uniform asymptoticity.

Campostrini and Rossi ignore completely point (2) above, which provides a concrete illustration that in actual fact, in $2D$, the $1/N$ expansion is non-uniform, at least for certain observables. They are also misrepresenting the numerical evidence, which shows systematically that the correlation length and the magnetic susceptibility grow faster than expected from asymptotic freedom (see Apostolakis et al [12]).

In conclusion, the mathematically well defined question is whether in the lattice $O(N)$ model the $1/N$ expansion can be used to deduce the correct behavior for $\tilde{\beta} \to \infty$ by looking at the large β behavior of finitely many terms. None of the observations made by Campostrini and Rossi in their Comment prove that point. To show that there is indeed a problem, we considered the ratio $G(\xi_\infty/2)/G(\xi_\infty/4)$, which for $\tilde{\beta} \to \infty$ approaches a finite value in the spherical model; we found that the coefficient of the $1/N$ correction to this ratio increases linearly with $\tilde{\beta}$ in $2D$, while it remains finite in $1D$. This fact casts doubt upon prior attempts to prove that $2D$ $O(N)$, $N > 2$ models have a mass gap for any $\beta < \infty$ via the $1/N$ expansion.

References

[1] M.Campostrini and P.Rossi, *Riv.Nuovo Cim.* **16** (1993) 1

[2] A.Kupiainen, *Commun.Math.Phys.* **73** (1980) 273

[3] P.Hasenfratz, *Phys.Lett.* **141B** (1984) 385

[4] J.Fröhlich, A.Mardin and V.Rivasseau, *Commun.Math.Phys.* **86** (1982) 87

[5] V.F.Müller, T.Raddatz and W.Rühl, *Nucl.Phys.* **B251[FS13]** (1985) 212

[6] H.Flyvbjerg and S.Varsted, *Nucl.Phys.* **B344** (1990) 646; H.Flyvbjerg and F.Larsen, *Phys.Lett.* **B266** (1991) 92; H.Flyvbjerg and F.Larsen, *Phys.Lett.* **B266** (1991) 99;

[7] P.Biscari, M.Campostrini and P.Rossi, *Phys.Lett.* **B242**] (1990) 225

[8] A.Patrascioiu and E.Seiler, *Superinstantons and the Reliability of Perturbation Theory in Non-Abelian Models* preprint MPI-Ph/93-87 and AZPH-TH/93-33

[9] A. Patrascioiu, AZPH-TH/91-49

[10] A.Patrascioiu and E.Seiler, *Nucl.Phys.(Proc.Suppl.* **B30** (1993) 184

[11] R.Gupta and C.F.Baillie, *Phys.Rev.* **B45** (1992) 2883

[12] J.Apostolakis, C.F.Baillie and G.F.Fox, *Phys.Rev.* **D43** (1991) 2687

[13] R.G.Edwards, S.J.Ferreira, J.Goodman and A.D.Sokal, *Nucl.Phys.* **B380** (1992) 621

[14] U.Wolff, *Phys.Lett.* **B248** (1990) 335

ANALYTIC DESCRIPTION OF THE MOTION OF A TRAPPED ION IN AN EVEN OR ODD COHERENT STATE

MICHAEL MARTIN NIETO[a]
Theoretical Division, Los Alamos National Laboratory
University of California, Los Alamos, New Mexico 87545, U.S.A.

and

Abteilung für Quantenphysik, Universität Ulm
D-89069 Ulm, GERMANY

A completely analytic description is given of the motion of a trapped ion which is in either an even or an odd coherent state. Comparison to recent theoretical and experimental work is made.

Recently, Matos Filho and Vogel[1] gave an analysis of a trapped ion[2] which, to very high precision, is in an even (+) or odd (-) coherent state[3]:

$$\psi_+ = [\cosh|\alpha|^2]^{-1/2}\sum_{n=0}^{\infty}\frac{\alpha^{2n}}{\sqrt{(2n)!}}|2n\rangle \ , \tag{1}$$

$$\psi_- = [\sinh|\alpha|^2]^{-1/2}\sum_{n=0}^{\infty}\frac{\alpha^{2n+1}}{\sqrt{(2n+1)!}}|2n+1\rangle \ . \tag{2}$$

Specifically, they gave lovely three-dimensional numerical graphs of the probability densities and Wigner functions, for the even and odd cases, as functions of position and time, for particular values of $\alpha = \alpha_1 + i\alpha_2$.

Here it is noted that closed-form expressions can be given for these wave functions, and hence for the probability densities and the Wigner functions. A direct way to evaluate the above sums[4] is to use[5] generating function techniques, yielding

$$\psi_+ = \left[\frac{e^{-\alpha^2}}{\pi^{1/2}\cosh|\alpha|^2}\right]^{1/2} e^{-x^2/2}\cosh(\sqrt{2}\alpha x) \ , \tag{3}$$

$$\psi_- = \left[\frac{e^{-\alpha^2}}{\pi^{1/2}\sinh|\alpha|^2}\right]^{1/2} e^{-x^2/2}\sinh(\sqrt{2}\alpha x) \ . \tag{4}$$

[a] Email: mmn@pion.lanl.gov

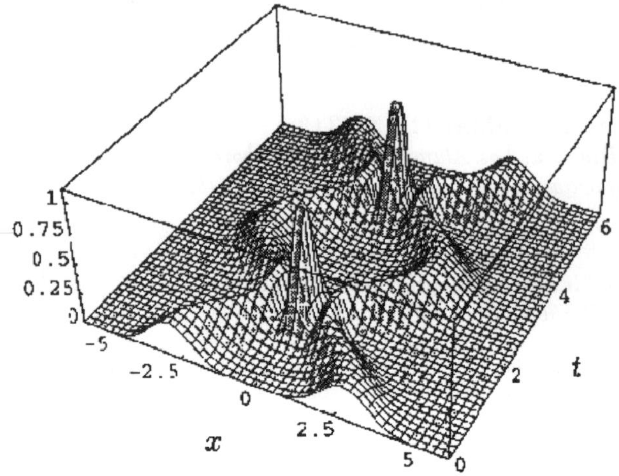

Figure 1: A three-dimensional plot of the even-coherent-state probability density, ρ_+, as a function of position, x, and time, t, for $\alpha_0 = 2$.

(One could, of course, obtain these expressions by other methods [3,6,7].) Physical intuition is satisfied when the above expressions are transformed to two Gaussians displaced on opposite sides of the origin [7]. Ignoring $e^{-i4\alpha_1\alpha_2}$,

$$\psi_\pm = \left[2\pi^{1/2}(1 \pm e^{-2|\alpha|^2})\right]^{-1/2} \left[e^{-(x-\sqrt{2}\alpha_1)^2/2+i\sqrt{2}\alpha_2 x} \pm e^{-(x+\sqrt{2}\alpha_1)^2/2-i\sqrt{2}\alpha_2 x}\right] . \quad (5)$$

Time-dependence can be included by letting $\alpha \to \alpha \exp[-i\omega t]$. We now take the convention $\alpha \to \alpha_0$ is real, as was done in Ref. 1. Then it is straighforward to calculate additional quantities. For example, the probability densities are

$$\rho_+ = \frac{e^{\alpha_0^2[\sin^2 \omega t - \cos^2 \omega t]}}{\pi^{1/2}[e^{\alpha_0^2} + e^{-\alpha_0^2}]} e^{-x^2}[\cosh\{2\sqrt{2}\alpha_0(\cos\omega t)x\} + \cos\{2\sqrt{2}\alpha_0(\sin\omega t)x\}], \quad (6)$$

$$\rho_- = \frac{e^{\alpha_0^2[\sin^2 \omega t - \cos^2 \omega t]}}{\pi^{1/2}[e^{\alpha_0^2} - e^{-\alpha_0^2}]} e^{-x^2}[\cosh\{2\sqrt{2}\alpha_0(\cos\omega t)x\} - \cos\{2\sqrt{2}\alpha_0(\sin\omega t)x\}]. \quad (7)$$

The above functions ρ_+ and ρ_+ describe the forms of Figs. 1 and 4 in Ref. 1. We show them in our Figs 1 and 2. Note in particular that the terms

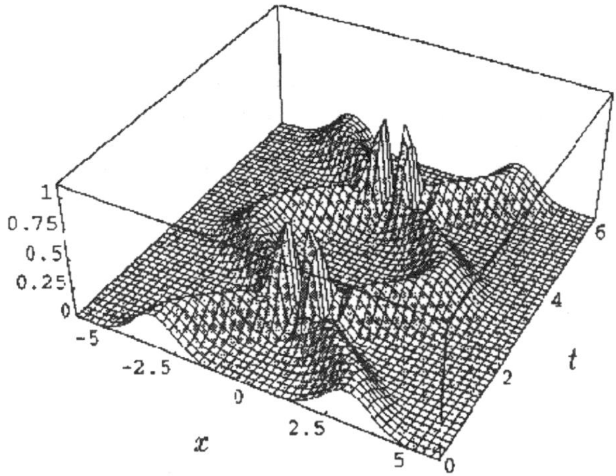

Figure 2: A three-dimensional plot of the odd-coherent-state probability density, ρ_-, as a function of position, x, and time, t, for $\alpha_0 = 5^{1/2}$.

$\exp[-x^2] \times \cosh$ describe the two "wave-packets" on opposite sides of the origin. The cos terms describe the interference effects near $x = 0$ at $t = (2j+1)\pi/2$.

These even and odd coherent states have recently been created by Wineland's group [2]. They took a trapped $^9Be^+$ ion and first cooled it to its zero-point energy. Then they created the even/odd states by a series of laser pulses that entangled the electronic and motional states of the ion. The separate packets were separated by as much as 800 Å with their individual sizes at 70 Å.

This same group has produced squeezed ground states [8]. Therefore, it is possible that squeezed even and odd states [9] may be produced in the future.

I gratefully acknowledge conversations held at the Humboldt Foundation Workshop on Current Problems in Quantum Optics, organized by Prof. Harry Paul. This work was supported by the U.S. Department of Energy and the Alexander von Humboldt Foundation.

References

1. R. L. de Matos Filho and W. Vogel, *Phys. Rev. Lett.* **76**, 608 (1996).
2. C. Monroe, D. M. Meekhof, B. E. King, and D. J. Wineland, *Science* **272**, 1131 (1996).
3. V. V. Dodonov, I. A. Malkin, and V. I. Man'ko, *Physica* **72**, 597 (1974).

4. M. M. Nieto and D. R. Truax, *Phys. Rev. Lett.* **71**, 2843 (1993).
5. M. M. Nieto and D. R. Truax, *Phys. Lett. A* **208**, 8 (1995).
6. M. Schubert and W. Vogel, *Phys. Lett. A* **68**, 321 (1978).
7. Other "two-packet" states are described in W. Schleich, M. Pernigo, and Fam Le Kien, *Phys. Rev. A* **44**, 2172 (1991).
8. D. M. Meekhof, C. Monroe, B. E. King, W. M. Itano, and D. J. Wineland, *Phys. Rev. Lett.* **76**, 1796 (1996).
9. M. M. Nieto, *Phys. Lett. A* **219**, 180 (1996).

CONTINUOUS STRUCTURES FOR THE DISCRETE SYMMETRIES

A. LUDU
Louisiana State University, Dept. Physics and Astronomy, Baton Rouge, LA 70803, U.S.A.

W. GREINER
Institut für Theoretische Physik der J. W. Goethe - Universität, D-60054, Frankfurt am Main, Germany

The non-Lie symmetries are mapped onto a space of differential transformations, defined by quasi-continuous q-deformations of dilations. This topological unification of discrete and continuous transformations yields a non-commutative structure of space-time. The q-deformation of the inversion symmetry produces two important effects: a fractional statistics and knotlike phase trajectories of soliton dynamics.

The continuous symmetry transformations are generated as smooth unfoldings of the infinitesimal transformations and have a special advantage over the discrete ones, the reason being the existence of infinitesimal generators, Greiner [1], Bohr [2]. The discrete symmetries have either a Z_2-graded structure (parity, reflections, mirroring, charge conjugation and time reversal) acting on one-particle states or on the space-time, or a permutation or braidlike structure, Lerda [3], acting on many-particle states.

The purpose of this contribution is to show that discrete symmetries can be continuously mapped toward unity, in the frame of the theory of q-deformed algebras, Chari and Pressley [4], and to present some physical consequences. Beyond the limiting cases of the deformations (which approach the original discrete symmetries) the new intermediate "nongeometric" symmetries bring new physical consequences: space-time discretization and non-commutative structure, phase-trajectories stability from knot theory and spontaneous breaking of symmetry, Ludu and Greiner [5]. To obtain this unification one has to introduce an associative continuous algebra of operators containing the discrete transformations among its elements.

If the discrete symmetries are smoothly connected it is possible to obtain new dynamical invariants from the Noether theorem. Simple examples of such continuous mapping of inversion are provided by the hydrodynamics of the soliton-antisoliton pair on shallow layers. Recently we introduced a new topological interpretation of the mirror transformation, Ludu and Greiner [5], by using representations of the quantum group $gl_q(2)$ in the space of the Reide-

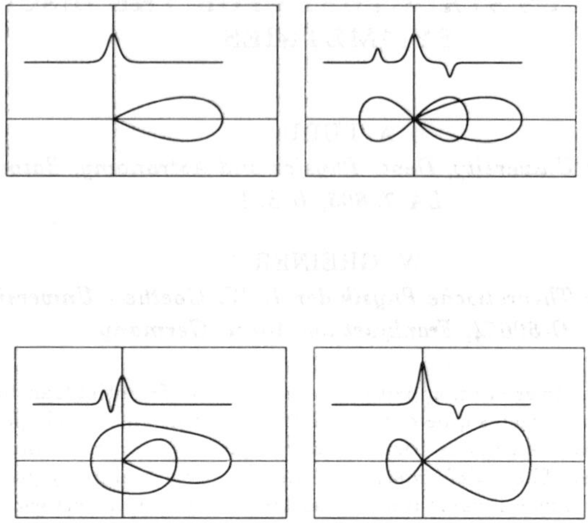

Figure 1: Unknoted one soliton solution (upper, left) compared with a soliton-antisoliton pair plus a higher soliton (upper, right). A breather solution (lower,left) described by a 1-knot, and a soliton-antisoliton solution (lower,right) having a 1-knot structure, too, but with a reverse orientation.

meister moves, Kauffman [6]. Multi-soliton solutions of the Korteweg de Vries equation are stable in time and transparent with respect to their scattering. They are caracterized by an infinite countable set of integral invariants. In the phase-space portrait, these solutions yield phase trajectories topologically equivalent to knotlike structures. The invariants of the soliton solutions are related in this way with the topological invaiants of the knots. Figs. 1, 2 present such examples of phase trajectories (in coordinates $(f(x,t), df/dt$ at a given x), together with the corresponding solution, $f(x,t)$. Any N-soliton solution of the Korteweg de Vries (KdV), Modified KdV or Sine-Gordon non-linear equations, is represented by a N0-knot loop in the phase space. The knotlike structure of these loops (i.e. number and signature of the knots) is invariant in time.

The discrete-continuous symmetry is realized by introducing a quantum analogue of the coordinates $x \rightarrow \bar{x}$

$$\bar{x} = x(q^{2x\partial} - 1)^{1/2}((q+1)(x\partial + 1))^{-1/2}, \tag{1}$$

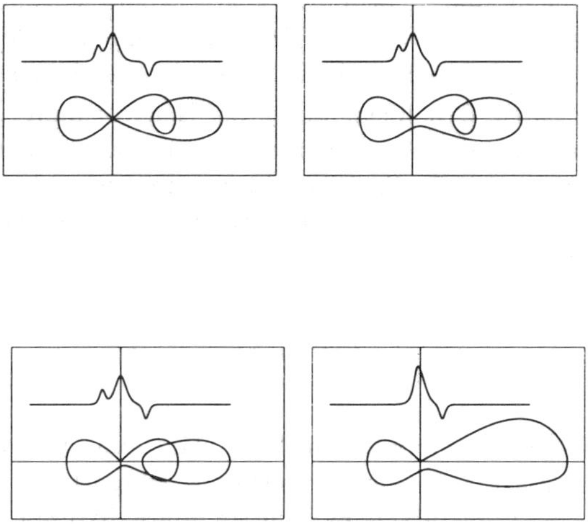

Figure 2: Three soliton solution in the scattering region produces a two knot structure (upper, left). A non-soliton similar solution has only one knot. Three soliton solutions generate 2-knot structure, too (lower).

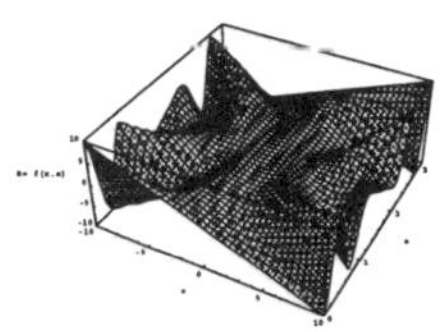

Figure 3: The continuous deformation of e^{x} into e^{-x}

fulfilling new q-deformed commutator relations, e.g. $[\bar{\partial}_x, \bar{x}]_q = 1$. In quantum mechanics it is widely used that to each linear operator not depending explicitely on time and commuting with the Hamiltonian it is possible to assign a constant of motion. For an arbitrary evolution equation of the form $i\partial_t \Psi = H(P)\Psi$ with $\Psi(x,t)$ a complex multicomponent function and H the Hamiltonian given by a Hermitian operator, the bilinear combination $I = \int \Psi^{\dagger} Q \Psi dx$, with Q a linear operator, is a constant of motion if $[i\partial_t - H, Q] = 0$. Consequently, new constants of motion can be found using new nongeometric symmetries. In the case of a one-dimensional example, this symmetry produces a gauge transformation of the potential, equivalent with a q-deformation of the unity operator

$$\tilde{V}(x,q^2) = \frac{q^2(q+q^{-1})}{2} q^{x\partial} \int \partial x V(x) d_q x, \tag{2}$$

that is the q-primitive [4] of the derivative of $V(x)$. For a multi-particle system, in a three-dimensional space, this results in a simpler support for the fractional statistics models. Rather than affixing by hand an internally prescripted vector potential $\vec{A}$ to an ordinary particle, which is then transmuted into an anyon, here, the apparition of $\vec{A}$ is the consequence of the non-commutativity of the space, Wess and Zumino [7]. In the non-commutative case the magnetic field is distributed on the plane, different from the localized magnetic field of anyons and cyons. A different statistics occurs, since the non-integrable phase-factor occuring in the wavefunction depend on the contour of integration. By introducing an uniform distribution of particles in the q-plane and assigning for each closed contour of integration a certain number of encircled particles, the transport of a certain particle along this closed loop is similar with a permutation between the external particles and those included in this contour. An application already occurs for systems of free electrons with quantum friction, on discrete time lattice, Ludu and Greiner [5].

Spontaneous symmetry breaking occurs for a free scalar real field $\phi(x,t)$, Coleman [8], when the mirror transformation of the space coordinate x or of the field ϕ is mapped to identiy. The problem is to find the most general Lagrangian invariant at discrete transformations, together with the its corresponding vacuum states. To obtain the symmetries of the Euler-Lagrange equation $\triangle(x,t,\phi,\phi_{tt},\phi_x,\phi_{xx}) = 0$ one identifies this equation with a manifold in the appropriate jet space. We introduce a vector field $v = a\partial_x + b\partial_\phi$ for the q-deformed dilation such that its second order prolongation $pr^2 v$ is tangent to the manifold, $pr^2 v \triangle|_{\triangle=0} = 0$. The most general form of Lagrangian invariant under the q-dilation, and consequently invariant under the mirror transformation $x \to -x$ is

$$L = \frac{1}{2}(\phi_t^2 + \phi_x^2 + (\phi_x - U(\phi))^2), \tag{3}$$

with $U(\phi)$ an arbitrary function. The invariance condition, reads

$$\frac{\partial U}{\partial \epsilon} + b\frac{\partial U}{\partial \phi} = b_x + (b_\phi - a_x)U - a_\phi U^2. \tag{4}$$

Testing the situation of exact symmetry, i.e. $b = 0$ (no modification of the vacuum solution) and $a = x$ (dilations in x) in Eq. 4 we obtain the trivial solution $\phi =$const. If we impose the freedom of having a spontaneous symmetry breaking, we add $b = \beta\phi$ (the most general dilation in ϕ) we obtain the most general form for the vacuum states:

$$\phi_x = f(\ln(\phi))\phi^{\beta-1}, \tag{5}$$

with β an arbitrary parameter. This equation, as resulted from the minimum condition imposed to L, is also the most general one for the real 1+1 case. If $\beta = 2$, one obtains the exponential solution $\phi_0 = const \cdot e^x$, which is invariant against inversion. Under the continuous inversion, this solution maps into $const \cdot e^{-x}$. The continuous transformation of this function is presented in Fig 3.

Acknowledgments

This work was supported by grants from the J. W. Goethe - Universität, Frankfurt am Main and the National Science Foundation.

References

1. W. Greiner and B. Müller, *Quantum Mechanics; Symmetries* (Springer-Verlag, Berlin, 1989).
2. A. Bohr and O. Ulfbeck, *Rev. Mod. Phys.* **67**, 1 (1995).
3. A. Lerda A and S. Sciuto S, NPB **401**, 613 (1993).
4. V. Chari and A. Pressley, *A Guide to Quantum Groups*, (Cambridge, University Press, 1994); A. Ludu and R. K. Gupta, JMP **34**, 5367 (1993).
5. A. Ludu and W. Greiner, *Found. Phys.*, in print, (q-alg/9612004);
6. L. H. Kauffman, *Knots and Physics*, (World Scientific, Singapore, 1991).
7. J. Wess and B. Zumino, preprint CERN-TH-5697/90, LAPP-TH-284/90.
8. S. Coleman, PRD **15**, 2929 (1977));

SOME QUESTIONS AND ANSWERS ABOUT CPT VIOLATION

V.A. KOSTELECKÝ
Physics Department, Indiana University,
Bloomington, IN 47405, U.S.A.

Minuscule violations of CPT and Lorentz invariance might arise in an extension of the standard model as suppressed effects from a more fundamental theory. In this contribution to the CarruthersFest, I present and answer some questions about CPT and the possibility of its violation.

1 Introduction

It is a pleasure for me to join in celebrating Pete's 61st birthday. In the spirit of Pete's approach to physics and to life, here are answers to some of the questions you always had about CPT and its violation (but were afraid to ask).

2 Basics

What is the CPT theorem?
The CPT theorem states that the product CPT of three discrete transformations, namely, charge conjugation C, parity reflection P, and time reversal T, is an exact symmetry of local relativistic field theories of point particles.[1] The theorem has withstood numerous high-precision experimental tests.[2] Indeed, CPT remains to date the only combination of C, P, T that is observed as an exact symmetry of nature.

Why consider CPT violation if the theorem says it isn't broken?
The CPT theorem is a general result holding for relativistic particle theories, and it can be experimentally tested to great accuracy. These facts make CPT violation an excellent candidate signature for unconventional physics, such as might arise in a fundamental theory based on extended objects like, say, strings.[3–5]

How could CPT be violated?
Theories disobeying any of the assumptions that enter the CPT theorem could violate CPT. In fact, one of the earliest explicit examples of CPT violation was presented by Pete.[6] It is a class of particle theories that naively appear

normal but that entail nonlocal interactions, excluded by assumption in the CPT theorem.

It is also possible to produce theories violating CPT for relatively subtle technical reasons. For example, the CPT theorem assumes that fields appear in finite-dimensional representations of the Lorentz group. Certain theories involving infinite-dimensional representations break CPT.[7] Also, since the CPT theorem holds within quantum field theory, it may be theoretically feasible to violate CPT if conventional quantum mechanics fails. This possibility has been suggested in the context of quantum gravity.[8]

Are there cases where CPT violation might be physically interesting?
A somewhat unexpected example where the CPT theorem does not appear to apply directly is ordinary quantum chromodynamics (QCD). The CPT theorem makes assumptions about the correspondence between the asymptotic Hilbert space and the fields in the theory that are open to question in QCD because of confinement. Rob Potting and I thought about this issue back in 1990 during our investigations of CPT symmetry in strings, but we didn't find an explicit proof or refutation of CPT invariance in QCD. To my knowledge, this aspect of CPT in QCD (and other confining theories) remains an open issue at present.

A particularly interesting and conceivably physical situation is spontaneous CPT violation.[3,4] In this case the dynamics of the action remains CPT invariant, which means many desirable properties of the theory are preserved. The violation occurs spontaneously in the solutions of the equations of motion, like the spontaneous breaking of the electroweak gauge group in the standard model. This type of CPT violation is a possibility in string theory, where the usual axioms of the CPT theorem may be modified because strings are extended objects.

3 Spontaneous CPT Violation

How could spontaneous CPT violation occur?
Suppose a higher-dimensional action that is Lorentz- and CPT-invariant underlies nature. The higher-dimensional Lorentz group would presumably be spontaneously broken by the solution to this theory, since it must represent our apparently four-dimensional world. This may induce spontaneous CPT breaking.

As an example, strings naturally exist in higher dimensions. Spontaneous Lorentz violation[9] is possible in string theory because string interactions exist that can trigger nonzero expectation values for Lorentz-tensor fields. Comparable interactions don't appear in conventional four-dimensional renormalizable gauge theories. If one or more of these tensors has an odd number of spacetime indices, CPT is also spontaneously broken.[3,4]

Can these ideas be verified explicitly in string theory?
For the field theory of the open bosonic string, the explicit action and equations of motion can be derived analytically for particle fields below some fixed level number N. Solutions have been found and compared for different N, in some circumstances to a depth of over 20,000 terms in the static potential.[10,11] These solutions include ones spontaneously breaking Lorentz and CPT invariance that persist as N is increased.

Does spontaneous CPT violation have to come from a string theory?
If spontaneous CPT violation emerges within a higher-dimensional theory then a string origin would presently seem to be the only possibility, since to my knowledge no other consistent candidate theories exist.

Does spontaneous CPT violation imply Lorentz violation or vice versa?
If the spontaneous CPT breaking arises from nonzero expectations of Lorentz tensors, then Lorentz invariance is necessarily spontaneously violated too. However, the converse is false, because expectation values of Lorentz tensors with an even number of indices preserve CPT.

Does spontaneous Lorentz violation imply causality is destroyed?
To my knowledge, there are no theoretical (or experimental) reasons to exclude (small) *spontaneous* Lorentz violation. Unlike other kinds of Lorentz breaking that do violence to accepted notions, spontaneous breaking is merely a feature of the solutions to the theory. The underlying dynamics remains Lorentz invariant. Indeed, it is possible to verify explicitly that microcausality is preserved in certain simple models arising from spontaneous Lorentz breaking.[12]

Other considerations make it seem very unlikely that a fundamental problem exists with the notion of spontaneous Lorentz violation. For example, the physics of a particle moving inside a biaxial crystal need not be (rotation or boost) Lorentz covariant, but this is merely a reflection of the presence of the background crystal fields and does not affect causality. Nonzero Lorentz-tensor expectation values throughout spacetime are similar in some respects and therefore also might be expected to have benign effects.

If spontaneous CPT/Lorentz breaking occurs, where are the Goldstone bosons?
Goldstone's theorem does not apply to discrete symmetries like CPT. If Lorentz invariance is treated as a global symmetry, its spontaneous breaking would indeed produce massless excitations, carrying quantum numbers related to the graviton. However, Lorentz invariance is believed to be local. In vector gauge

theories, Goldstone bosons would be absorbed by the gauge fields, which become massive through the Higgs mechanism. In the present case, the graviton propagator is affected but no graviton mass is generated.[9]

4 Standard-Model Extension

If spontaneous CPT breaking occurs in higher dimensions, is it observable?
If the mechanism of spontaneous Lorentz and CPT violation occurs in a higher-dimensional theory, it would seem likely to involve the four physical dimensions too. However, as neither Lorentz nor CPT breaking have been experimentally observed, any effects at the level of the standard model must be highly suppressed.[3–5]

What would be the scale of the suppression?
Taking the scale governing the fundamental theory as the Planck mass $m_{\rm Pl}$ and denoting the electroweak scale by $m_{\rm ew}$, the natural suppression factor for Planck-scale effects in the standard model is[3–5] $m_{\rm ew}/m_{\rm Pl} \simeq 10^{-17}$. A factor this small means that only a few Lorentz and CPT-violating effects are likely to be observable.

How would effects in the fundamental theory appear in a low-energy theory?
The fermionic sector of the four-dimensional low-energy effective theory might, for example, contain terms of the form[4,5]

$$\mathcal{L} \sim \frac{\lambda}{M^k}\langle T\rangle \cdot \overline{\psi}\Gamma(i\partial)^k\chi + h.c. \quad . \tag{1}$$

Here, a fermion bilinear involving a gamma-matrix structure Γ and derivatives $i\partial$ is coupled to the expectation value of a Lorentz tensor T, which breaks Lorentz and CPT symmetry. The coupling coefficient involves a dimensionless coupling constant λ and an appropriate power of some large scale M, such as the Planck or compatification scale.

Is there an extension of the standard model that includes these effects?
A general extension of the standard model, including Lorentz-breaking terms both with and without CPT violation, has been obtained.[12] The extra terms maintain the usual SU(3) × SU(2) × U(1) gauge invariance and are power-counting renormalizable. A framework has also been given for treating theoretically the effects of spontaneous CPT and Lorentz breaking.

5 Experimental Tests

How can CPT be tested experimentally to high precision?
Oscillations of neutral mesons P, where P is one of K, D, B_d, or B_s, are sensitive probes of CPT violation by virtue of their interferometric nature.[5,13–15] The time evolution of the oscillations is governed by a 2×2 effective hamiltonian Λ. Conventional quantum mechanics allows in principle two complex CP-violating parameters to appear in Λ: the usual CP- and T-violating parameter ϵ_P that preserves CPT, and a CP- and CPT-violating parameter δ_P that preserves T. Experiments bounding the value of δ_P can test CPT to high precision.

How do the theoretical modifications affect experimental observables?
Within the CPT-violating extension of the standard model, nonzero values of δ_P emerge from small corrections to conventional perturbative calculations. For a given P system, it turns out that δ_P is given by[4,5]

$$\delta_P = i \frac{h_{q_1} - h_{q_2}}{\sqrt{\Delta m^2 + \Delta\gamma^2/4}} e^{i\hat{\phi}} \quad . \tag{2}$$

Here, the experimental observables Δm and $\Delta\gamma$ are mass and rate differences, with $\hat{\phi} = \tan^{-1}(2\Delta m/\Delta\gamma)$. The parameters $h_{q_j} = r_{q_j}\lambda_{q_j}\langle T \rangle$ are determined by coefficients of terms in the standard-model extension and by factors r_{q_j} from the quark-gluon sea.

Are there definite signals from spontaneous CPT violation?
Assuming hermiticity of the standard-model extension, the h_{q_j} are real. This implies the condition[4,5]

$$\mathrm{Im}\, \delta_P = \pm \frac{\Delta\gamma}{2\Delta m} \, \mathrm{Re}\, \delta_P \quad . \tag{3}$$

Moreover, the severity of the suppression factor for Planck-scale effects suggests direct CPT violation in P-meson decay amplitudes is unobservable. The relation (3) for indirect CPT violation and the absence of direct CPT violation are signatures for spontaneous CPT violation in any P system.

In addition, the CPT-violating couplings in the standard-model extension seem likely to differ substantially for distinct quarks, as do the Yukawa couplings. The CPT-violating quantities δ_P could therefore vary significantly for different P. This means CPT should be tested in more than one neutral-meson system. Given the sparsity of present bounds on CP violation in the B_d system, it is even possible that $|\delta_{B_d}| > |\epsilon_{B_d}|$, in which case CPT effects would dominate conventional CP ones in the proposed B factories.

What are the current limits and prospects for future tests?
The kaon system offers the best CPT bound from neutral mesons. The published limits[2,16] on $|\delta_K|$ are of order 10^{-3}. Completed experiments (e.g., CPLEAR at CERN), ongoing ones (e.g., KTeV at Fermilab), and ones currently being designed are likely to improve the bounds in the near future.

Mixing has not yet been seen in the D system, and dispersive effects make theoretical predictions uncertain. In favorable conditions some tests of CPT symmetry could be feasible,[5,13] perhaps even with current data and probably with statistics available within the next decade.

The B_d system might involve the largest CPT violation because it includes the heavy b quark. Enough data to bound δ_{B_d} at the level of order 10% have already been obtained in the CERN LEP experiments and in CLEO experiments at Cornell.[5,14,15] Indeed, the OPAL collaboration at CERN has very recently placed a bound[17] on $\mathrm{Im}\,\delta_{B_d}$ of about 2×10^{-2}. The many B-dedicated experiments now being developed are likely to improve this bound considerably.

Are there any tests in systems other than neutral mesons?
Several possibilities exist, including signals that might emerge from the CPT-violating extension of quantum electrodynamics implied by the standard-model extension.[12,18] For example, CPT violation can potentially be tightly constrained by experiments establishing the difference between the electron and positron anomalous magnetic moments. Bounds could be placed on leptonic parameters for CPT violation that are comparable to those in neutral mesons. Further bounds may emerge from photon properties. It is important to consider a variety of tests because the standard-model extension allows distinct parameters to control effects in the different sectors.

What does CPT violation imply about the observed baryon asymmetry?
Conventional baryogenesis requires nonequilibrium processes and C- and CP-breaking interactions.[19] However, an acceptable mechanism for baryogenesis in thermal equilibrium might emerge from terms of the form (1) under suitable conditions.[20] A large asymmetry could be produced at grand-unification scales, subsequently being diluted to the observed value through sphaleron or other effects.

Acknowledgments

I thank Orfeu Bertolami, Robert Bluhm, Don Colladay, Rob Potting, Neil Russell, Stuart Samuel, and Rick Van Kooten for collaborations. This work was supported in part by the United States Department of Energy under grant number DE-FG02-91ER40661.

References

1. See, for example, R.F. Streater and A.S. Wightman, *PCT, Spin and Statistics, and All That,* Benjamin Cummings, Reading, 1964.
2. See, for example, R.M. Barnett *et al.*, Review of Particle Properties, Phys. Rev. D **54** (1996) 1.
3. V.A. Kostelecký and R. Potting, Nucl. Phys. B **359** (1991) 545.
4. V.A. Kostelecký, R. Potting, and S. Samuel, in S. Hegarty et al., eds., *Proceedings of the 1991 Joint International Lepton-Photon Symposium and Europhysics Conference on High Energy Physics,* World Scientific, Singapore, 1992;
   V.A. Kostelecký and R. Potting, in D.B. Cline, ed., *Gamma Ray–Neutrino Cosmology and Planck Scale Physics* (World Scientific, Singapore, 1993) (hep-th/9211116).
5. V.A. Kostelecký and R. Potting, Phys. Rev. D **51** (1995) 3923.
6. P. Carruthers, Phys. Rev. Lett. **18** (1967) 353; Phys. Lett. **B62** (1968) 158; J. Math. Phys. **9** (1968) 928, 1835; Phys. Rev. **172** (1968) 1406.
7. A.I. Oksak and I.T. Todorov, Commun. Math. Phys. **11** (1968) 125.
8. S.W. Hawking, Phys. Rev. D **14** (1976) 2460.
9. V.A. Kostelecký and S. Samuel, Phys. Rev. D **39** (1989) 683; *ibid.,* **40** (1989) 1886; Phys. Rev. Lett. **63** (1989) 224; *ibid.,* **66** (1991) 1811.
10. V.A. Kostelecký and S. Samuel, Nucl. Phys. B **336** (1990) 263; Phys. Rev. Lett. **64** (1990) 2238; Phys. Rev. D **42** (1990) 1289.
11. V.A. Kostelecký and R. Potting, Phys. Lett. B **381** (1996) 389.
12. D. Colladay and V.A. Kostelecký, Phys. Rev. D **55** (1997) 6760; Indiana University preprint IUHET 359.
13. D. Colladay and V.A. Kostelecký, Phys. Rev. D **52** (1995) 6224.
14. D. Colladay and V.A. Kostelecký, Phys. Lett. B **344** (1995) 259.
15. V.A. Kostelecký and R. Van Kooten, Phys. Rev. D **54** (1996) 5585.
16. L.K. Gibbons et al., Phys. Rev. D **55** (1997) 6625.
17. OPAL Collaboration, R. Ackerstaff et al., CERN preprint CERN-PPE/97-036 (April 1997).
18. R. Bluhm, V.A. Kostelecký, and N. Russell, Indiana University preprint IUHET 358.
19. A.D. Sakharov, JETP Lett. **5** (1967) 24.
20. O. Bertolami, D. Colladay, V.A. Kostelecký, and R. Potting, Phys. Lett. B **395** (1997) 178.

DISSIPATION, ENTROPY, CHAOS, INFORMATION IN QUANTUM PHYSICS

Michael Danos*
Enrico Fermi Institute, University of Chicago
Chicago, Illinois 60637

Abstract

As a consequence of the Heisenberg uncertainty any quantum physics system inescapably undergoes relaxation towards thermal equilibrium, thus towards maximum entropy. The non-dissipative states have probability measure zero and can not exist in nature. The impact these results have on chaotic behavior of quantum systems will be discussed, as well as the difference between quantum and classical systems, which have vanishing uncertainty.

1 Introduction

The inevitability of dissipation in quantum physics has been demonstrated in a recent paper [1], with the result that even though in quantum physics a system is described by a Hamiltonian which is time-reversal invariant, the time-reversed state of any physical solution also is a solution, but that it cannot exist in nature in principle, having probability measure zero. All other physical states as consequence of the hermiticity of the Hamiltonian evolve towards equal occupation probability for the accessible phase space cells, i.e., towards thermal equilibrium. Replacing the classical probabilities by quantum probabilities in Boltzmann's expression for the entropy one immediately obtains a formula for the entropy in quantum physics. As long as thermal equilibrium has not been reached, the process of dissipation allows the definition of the arrow of time. However, when attempting to compute the evolution of a system the inevitable inaccuracies – which in general will by far exceed the Heisenberg uncertainties – will yield increasing entropy both for computing forward in time and for computing backward in time.

These results now allow an intercomparison between classical and quantum dissipation, which show that strictly dissipation is not compatible with classical physics. Also, classical deterministic chaos survives incorporation into quantum physics only by acquiring qualitative modifications. We also give a short discussion of thermodynamics and information.

2 Classical Dissipation

It is not clear that a consistent structure "Classical Physics" is possible, even leaving out atomic physics. Thus, for example, the coarse-graining of classical statistical thermodynamics, which consists in averaging the initial state over a finite volume of phase space, lies outside of the frame of classical physics: there a particle has definite position and momentum. Besides, it is no panacea: if the coarse-graining volume is taken as a compact manifold, then according to the Liouville theorem that volume is a constant of the motion, and hence the entropy will stay constant. Since classical physics does not have a Heisenberg-type uncertainty relation, one of the underlying reasons for the existence in quantum physics of dissipation is absent. This suggests that in classical physics dissipation in the strict sense may not apply.

Consider now a Gibbs ensemble. Given any starting point in the system phase space and given a small volume compact manifold including that starting point, then according to Poincaré's theorem there will be an infinite number of crossings of the system phase space trajectory through that volume, i.e., the system will experience an infinite number of quasi-returns. Since that volume can be chosen arbitrarily small, but not zero, then, in order to arrive in this volume, the system must re-do all the (classical dissipative!) evolution it had undergone before. Thus in classical physics there exists an infinite number of closed or almost closed trajectories with non-zero measure, for which the entropy repeatedly grows and decreases. Thus in that process the second law of thermodynamics must be massively violated. This is in contrast to quantum physics where no closed or quasi-closed trajectories exist[1].

The origin of this essential difference between quantum and classical evolution can be seen in the difference between the characteristics of the trajectories: classically it is a point trajectory, in quantum physics it is a wave trajectory. The former is continuous and one-dimensional; it can be equally traversed 'forward' and 'backward'. In contrast, the wave trajectory at each scattering branches into waves entering all channels; this is the reason the quantum trajectory disperses over the accessible phase space. It is a 'trapdoor' evolution in that the time-reversed motion is not possible, as discussed in Ref.[1]. Classical trajectories do not split up; the trajectories containing the indefinite point "to continue left or to continue right of an obstacle" have measure zero.

On the other hand, dissipation in a loose sense indeed can exist in classiscal physics: the probability that for an arbitrary phase space point the evolution (upon repeated coarse-graining, which breaks the frame of classical physics by imitating the superposition of states by means of the quantum physics density matrix) will be towards increasing entropy can be *assumed* to be overwhelming. Furthermore, the Poicaré returns can be safely ignored. This is the point of view generally taken as a starting point of the discussion of dissipation, also for quantum physics[1], which, however, in view of the present paper is not needed.

3 Quantum–Classical Chaos

The result that all systems in quantum physics ultimately approach thermal equilibrium in an exponential manner indicates that a behavior like that of a classical chaotic motion – where the trajectories stay in the vicinity of the strange attractor but do not reach it, maintaining a well-defined average distance[3] – strictly speaking does not occur in quantum physics. Thus classical-type chaos can occur only in those systems which allow the approximate description by quasi-classical degrees of freedom, which can exist only over limited periods of time, viz., only as long as these degrees of freedom are far from thermal equilibrium. The system U-matrix, which desribes the time evolution, in such systems is quasi-disconnected, allowing the quasi-classical degrees of freedom to remain isolated from the rest of the system for a sufficiently long time to be accessible to observation.

Since classically chaotic systems do exist, the wave functions (density matrices) of such systems will certainly reproduce that chaotic motion over a certain time period, i.e., before the quantum dissipation process takes over. During that time these systems will evolve in such a manner that owing to Ehrenfest's theorem the expectation values of the appropriate operators will exhibit the chaotic motion [4]. Without the classical chaotic behavior it would be extremely difficult to recognize this behavior in a purely quantum treatment, i.e., to recognize the relevant collective quasi-classical degrees of freedom. On the other hand, it should be quite feasible to go the other way. When trying to demonstrate that quasi-classical chaotic motion by writing down the, say, density matrices for such systems one only has to be certain that one has not written a zero measure quantity. This, however, is in not too serious since, unless one one uses the time-reversed of a physical state, it is not easy to write a purely zero measure quantity, and a zero measure component in an overall non-zero measure function in the evolution retains its zero measure and hence does not contribute to any experimentally testable predictions. In order to be sure that one has actually defined an "allowed" system [1] one only must be able to construct a gedanken-method for realizing the assumed initial conditions, of course not involving a time-reversal transformation.

An example of a quantum system of that quasi-chaotic kind has recently been given by Cao and Hwa [5], where the quasi-classical variables are the (in cloud chambers observable) particle trajectories resulting from the hadronization of a quantum object.

The subject denoted by "quantum chaos" in today's literature [6] concerns a different subject, namely, the characteristics of wave functions of systems or of ensembles of systems at energies close to the ground state. A particularly clear example is given in ref. [7] which shows the (standing wave) density pattern of a one-body sytem in a 2-dimensional Sinai stadium.

A rather far-fetched example of the interplay classical – quantum physics is the generic three-body problem, containing three point particles interacting with instantaneous gravitation. In classical physics a general proof of the instability of this system is not known. On the other hand, in "nature" this system would simply be a quantum system albeit perhaps having immensely large quantum numbers. Hence it will exhibit a gravitational Auger effect, and thus undoubtedly is unstable. It is not clear, however extremely likely, that this instability will transfer to a strictly classical dynamics.

4 Information

The use of the term "entropy" [8] or "negentropy" [9] to characterise the content of a message, also termed "information", coupled with the v. Neumann definition of the quantum entropy, has led to rather widespread confusion, in particular in the context of black hole thermodynamics. To wit, as mentioned above, the v. Neumann "entropy" is an exactly conserved quantity, which is not the case for the thermodynamic entropy [110]. However, the conclusion is drawn that since the "entropy" is conserved, so should be the "information".

Quantum physics indeed contains conserved quantities associated with zero restmass gauge particles (photons, gluons, gravitons): energy, momentum, angular momentum, electric and color charges. No conservation law exists for "information". Indeed, take a (coherent or incoherent) light source, and transmit a (finite length) string of Morse code. Even without any dissipative mechanism, owing to the $1/r^2$ law after some distance the message will be irretrievably lost in the cosmic thermal background radiation. Scattering from the residual gas of interstellar space will add to it further loss of "information" owing to the ineviable dissipation of that process.

Concerning the "loss of information" in gravitational collapse, this question must wait for the development of quantum gravity to allow resolution, in the sense of determination of which, if any, are the quantities beyond the presently recognized ones which will survive the collapse and the Hawking evaporation.

Of course, the field of "information entropy", being in fact disconnected from thermodynamic entropy, remains a very useful tool for analyzing data, be it from heavy ion collisions, or in communication engineering.

5 Conclusions

As shown in Ref. [1] the Heisenberg uncertainy relations forbid the existence of time-reversal invariant states, and as a consequence of the time-reversal invariance of the Hamiltonian quantum dynamics achieves directly relaxation towards a uniform occupation of the accessible phase space. This way, in quantum physics thermalization, i.e., dissipation, is an inevitable process; all systems tend towards thermal equilibrium, while exhibiting quantum fluctuations. As long as thermal equilibrium has not been reached, the process of dissipation allows the definition of the arrow of time.

Since the Heisenberg uncertainty principle does not apply, classical physics allows only for quasi-dissipation, which includes the massive violations of the second law associated with the Poincaré returns. To avoid these violations one must break the frame of classical physics in some manner, e.g., by coarse-graining. The classically possible eternal chaotic motion in quantum physics unavoidably dissipates into thermal equilibrium; hence it can have only a limited lifetime. Concerning "information", in contrast to the purity of a state no conservation law for it exist in quantum physics. It is at present unknown whether any other than the exactly conserved quantities can survive the collapse into a black hole.

References

* Visiting Scholar

1. M. Danos, Phys. Rev. E **52**, 3637 (1995)
2. U. Weiss, *Quantum Dissipative Systems*, World Scientific, Singapore (1992)
3. H. D. I. Abarbanel, R. Brown, J. J. Sidorowich, and L. Sh. Tsimring, Rev. Mod. Phys. **65**, 1331 (1993)
4. J. C. Robinson *et al*, Phys. Rev. Lett. **74**, 3963 (1995)
5. Z. Cao and R. C. Hwa, Phys. Rev. Lett. **75**, 1268 (1995).
6. A large number of papers on that subject have been, and are being, published. A recent example is A. V. Andreev, O. Agam, B. D. Simons, and B. L. Altshuler, Phys. Rev. Lett. **76**, 3947 (1996). For monographs see e.g., M.-J. Giannoni, A. Voros, and J. Zinn-Justin, it Chaos and Quantum Physics; North Holland, Amsterdam (1991); F. Haake, *Quantum Signatures of Chaos*; Springer-Verlag, Berlin (1991)
7. V. N. Prigodin, N. Taniguchi, A. Kudrolli. V. Kidambi, and S. Sridhar, Phys. Rev. Lett. **75**, 2392 (1995). See in particular Fig.1, which shows the pattern of the energy density in a microwave cavity. That, of course, is an open system: energy is being continouosly supplied to offset the dissipation in the cavity walls.
8. C. E. Shannon, *The Mathematical Theory of Communication*; Univ. Illinois Press (1949)
9. L. Brillouin, *Science and Information Theory*; Academic Press, New York (1962)
10. R. M. Wald, Phys. Rev. D **20**, 1271 (1979)

CHAOTIC BEHAVIOR OF QUARKS

RUDOLPH C. HWA
Institute of Theoretical Science and Department of Physics
University of Oregon, Eugene, OR 97403-5203, USA
E-mail: hwa@oregon.uoregon.edu

For the quark-gluon system the usual description of chaos that is appropriate for classical nonlinear systems is not applicable, so a new measure of chaos is needed. It is shown that the entropy indices μ_q, which characterize the fluctuations of spatial patterns, are as effective as the Lyapunov exponent in measuring chaos. When applied to the QCD parton showers, it is found that the quark jets are more chaotic than gluon jets. The analysis is highly appropriate for a quantitative treatment of the erratic fluctuations of multiparticle production in leptonic, hadronic and nuclear collision processes. It represents a step beyond the traditional intermittency analysis.

1 Introduction

The topic to be discussed here concerns the possibility of chaotic behaviors of a quark-gluon system. Since the number of degrees of freedom of such a system is not conserved, the traditional method of following the distance between two nearby trajectories of a classical system cannot be applied. Thus a major part of our effort in the problem is to identify an appropriate measure of chaos and to show that it is as good as the usual one when applied to the classical problems. With the new measure it is then possible to investigate the question of chaos for QCD parton showers. The significance of this line of investigation is in uncovering the fluctuating properties of multiparticle production that have hitherto been averaged over. Those fluctuations are in the event multiplicity and in the phase-space distribution of the particles produced. Are they unpredictable? In what way can they be recognized as chaotic behavior? The experimental determination of the key measures that quantify those fluctuation properties will then present a challenge to theoretical models, which may not have been sufficiently accurate to reproduce them.

In order to introduce a quantitative measure of unpredictability we shall proceed in stages. First, we shall discuss the notion of erraticity, which concerns the fluctuations of the spatial patterns of the final states in momentum space. Then we shall discuss entropy indices, followed by their applications to both the classical nonlinear problems and the QCD parton showers. The consideration involves an area of phenomenology that has hitherto been unexplored.

2 Erraticity and the Entropy Indices

Intermittency refers to the scaling behavior of the multiplicity fluctuations in bins of size δ.[1] In determining the normalized factorial moments by averaging over all events at a fixed bin and then averaging over all bins, information on the spatial patterns from event to event is lost. Erraticity analysis is an attempt to capture that information.

Let F_q denote a specific quantification of the spatial pattern in the phase space of a final state. It can be the horizontal factorial moments or the correlation integrals. After many events one obtains a distribution $P(F_q)$ of F_q. Now, let us define the normalized (vertical) moments [2]

$$C_{p,q} = \langle F_q^p \rangle \, / \, \langle F_q \rangle^p \quad , \tag{1}$$

where the averages are done with $P(F_q)$ as the probability distribution. The fluctuations of F_q become more simple to categorize and more interesting to study, if $C_{p,q}$ exhibits scaling behavior, i.e.,

$$C_{p,q} \propto M^{\psi_q(p)} \quad , \tag{2}$$

where M is the number of bins in a fixed space. For convenience, we refer to this behavior as erraticity.[3] It is a natural extension of the notion of intermittency. The value of p can be any positive real number. It should not be negative because F_q may vanish for some events, so $P(F_q = 0) \neq 0$ for some q. For $0 < p < 1$, it is the $F_q < 1$ region that is probed by $\psi_q(p)$, while for $p > 1$ it is the spiky events with high F_q that $\psi_q(p)$ describes. In practice it is not necessary to consider p greater than 3.

The entropy indices are defined by

$$\mu_q = \left. \frac{d}{dp} \psi_q(p) \right|_{p=1} \quad . \tag{3}$$

We refer to μ_q as the entropy indices.[2] A direct way of calculating μ_q is to write $C_{p,q}$ as

$$C_{p,q} = \langle \Phi_q^p \rangle \quad , \qquad \Phi_q = \frac{F_q}{\langle F_q \rangle} \quad , \tag{4}$$

so that

$$\left. \frac{d}{dp} C_{p,q} \right|_{p=1} = \langle \Phi_q \ln \Phi_q \rangle \quad . \tag{5}$$

If $C_{p,q}$ has the scaling behavior (2), then we also have

$$\left. \frac{d}{dp} C_{p,q} \right|_{p=1} = \mu_q \ln M \tag{6}$$

in the scaling range of M. Consequently, we obtain

$$\mu_q = \frac{\partial}{\partial \ln M} \langle \Phi_q \ln \Phi_q \rangle \quad , \tag{7}$$

which is to be determined in the region where $\Sigma_q \equiv \langle \Phi_q \ln \Phi_q \rangle$ exhibits a linear dependence on $\ln M$.

3 Chaoticity

We now consider the question whether multiparticle production processes are chaotic. Let us first recall the properties of chaotic behavior in classical nonlinear dynamics.[4] Since in such systems trajectories in space-time exist, the distance function $d(t)$ between two trajectories can be defined. A system is chaotic if $d(t) \sim e^{\lambda t}$, $\lambda > 0$, no matter how small $d(0) = \epsilon$ may be, for $\epsilon > 0$. For classical Yang-Mills dynamics the Lyapunov exponent λ has been shown to be positive by lattice calculation.[5]

For quantized Yang-Mills fields the problem is vastly more complicated. In addition to the ambiguity associated with quantum chaos in the realm of first quantization, we now have also the problems of nonconservation of the number of degrees of freedom and of the lack of a meaningful definition of trajectory. The absence of an unambiguous notion of time in production processes further results in the nonexistence of the Lyapunov exponent for the problem.

There are two issues here. One is to find a measure of chaos, and the other concerns QCD branching processes. We discuss them separately.

3.1 A new measure of chaos

Our proposal is to use μ_q as an alternative measure of chaos in problems where only the spatial patterns can be observed. Whether that approach agrees with the conventional approach can be determined only by applying the measure to known chaotic systems, of which there are many with simple nonlinear dynamics.[4] They are classical systems with well-defined trajectories, and the positivity of the maximum Lyapunov exponent λ is a sufficient criterion for chaos. Since our measure is concerned with spatial patterns rather than temporal evolution, it is necessary to construct a spatial pattern for every trajectory. That is achieved by considering a set of points corresponding to the positions of a trajectory at a discrete collection of times, separated by finite intervals apart. Clearly, two nearby trajectories for $\lambda < 0$ would generate two very similar patterns, while two chaotic trajectories for $\lambda > 0$ would have very dissimilar patterns. We can generate $\mathcal{N}$ events by considering $\mathcal{N}$ trajectories all starting from a small neighborhood of an initial point. Studying the $P(F_q)$ distribution of those events and then comparing the resultant μ_q to λ constitute our procedure to check the effectiveness of μ_q as a measure of chaos.

We have applied this procedure to the logistic map and the Lorenz attractor, the details of which are omitted here. They are nonlinear systems that become chaotic when their control parameters r exceed certain critical values.[4] We present here only the result for the logistic map.[6] In Fig. 1 the dashed line shows the value of λ as a function of r. For $r > 3.57$, λ becomes positive, although there are short intervals where λ drops below zero. The solid line indicates the value of μ_2 (normalized by a specific factor), calculated at discrete points in the same range of r. Evidently μ_2 coincides very well with λ except that it cannot be negative but vanishes when $\lambda \leq 0$. The positivity of λ is coordinated with the positivity of μ_2 so that the fluctuations of spatial patterns can equally serve as a means for revealing chaotic behaviors.

3.2 Chaos in QCD parton showers

Since perturbative QCD is well developed theoretically, and the final states of quark and gluon jets are well measured experimentally, we propose that the erraticity of hard processes be used as a new arena for comprehensive tests and that the entropy indices of parton showers be determined to reveal the chaoticity of QCD.

Since hadronization is irrelevant to the question of chaos in pQCD, we have developed an efficient Monte Carlo generator of quark and gluon jets with the usual Sudakov form factor and splitting functions incorporated in the code.[2] The initial virtuality of a quark (Q jet) or a gluon (G jet) is fixed at a Q^2 value for all events, and all partons evolve and branch until the virtuality of a parton reaches $\leq Q_0^2$, where branching is terminated for that parton. From the spatial pattern of each event we calculate the horizontal normalized factorial moments F_q, then the distribution $P(F_q)$, the vertical moments $C_{p,q}$, the erraticity indices $\psi_q(p)$, and finally the entropy indices μ_q. The results of the calculation for $Q/Q_0 = 10^3$ are shown in Fig. 2. The positivity of μ_q for both Q and G jets indicates that the QCD branching processes are chaotic. Moreover, the Q jets are more chaotic than the G jets. The reason is that quark jets have fewer partons produced, so the multiplicity fluctuations relative to the mean is larger.

Using the positivity of μ_q as a criterion for chaos, we have not only found that the non-Abelian, nonlinear dynamics of QCD is chaotic, but also come to realize that any quantum system in which the number of quanta of interest is not conserved is likely to have positive μ_q. Thus while chaotic behaviors are spectacular and remarkable phenomena in classical dynamics, their occurrences seem to be common and generic in quantum systems. Perhaps the notion of chaos should not be more emphasized in multiparticle production than the realization that the unpredictability of the final states can be quantified. To calculate the entropy indices for all collision processes, soft as well as hard, that can agree with the experimental data would be a theoretical challenge. Our results obtained should not be taken quantitatively for comparison with experiments, since the MC code has not been tuned to check the other features of parton showers. However, they have been effective in demonstrating the procedure of determining μ_q and in elucidating the question of chaos in QCD.

Acknowledgments

The research underlying the contents of this talk was done in collaboration with Z. Cao. It was supported, in part, by the U.S. Department of Energy under Grant No. DE-FG03-96ER40972.

1. A. Białas and R. Peschanski, *Nucl. Phys.* B **273**, 703 (1986); **308**, 857 (1988).
2. Z. Cao and R. C. Hwa, *Phys. Rev. Lett.* **75**, 1268 (1995); *Phys. Rev.* D **53**, 6608 (1996); **54**, 6674 (1996).
3. R C. Hwa, *Acta. Phys. Polon.* **B27**, 1789 (1996).
4. H. G. Schuster, *Deterministic Chaos*, 3rd ed. (Physik-Verlag, Weinheim, 1995).
5. B. Müller and A. Trayanov, *Phys. Rev. Lett.* **68**, 3387 (1992); T. S. Biró, C. Gong, B. Müller and A. Trayanov, *Int. J. Mod. Phys.* **C5**, 113 (1994); C. Gong, *Phys. Lett.* B **298**, 257 (1993), *Phys. Rev.* D **49**, 2642 (1994).
6. Z. Cao and R. C. Hwa, OITS-610 (submitted to Phys. Rev. E).

Turbulent Cascades

Martin Greiner[a], Peter Lipa[b] and Jens Giesemann[a+c]

[a] *Institut für Theoretische Physik, Technische Universität, D-01062 Dresden, Germany*
[b] *Institut für Hochenergiephysik der Österreichischen Akademie der Wissenschaften, Nikolsdorfergasse 18, A-1050 Vienna, Austria*
[c] *Institut für Theoretische Physik, Justus Liebig Universität, D-35392 Gießen, Germany*

Discrete cascade models for fully developed turbulence are, due to their binary hierarchical structure, not translationally invariant; the implications on the deviation from the underlying selfsimilarity is discussed. We show that wavelet bases represent approximate normal coordinates for these binary cascade processes and simplify their study. More general processes with overlapping branches can be efficiently modeled and analyzed via wavelet cascades.

Having had the privilege of working as postdocs with Peter Carruthers, let us tell you how he shaped and directed once again the future thinking of two young scientific greenhorns (MG and PL). One day he called us into his office and welcomed us with a warm smile: "Have you ever heard about wavelets?". Of course we hadn't. "Well", he continued, "it smells good. Look into it!". This short advice has influenced our research for the past five years and probably will do so for many more years to come.

Soon after Pete's initial push we were convinced of the power and usefulness of wavelets and started applying them to the study of cascade models. Today it is obvious that the "fractals boom" of the eighties is being replaced by a comparable "wavelet boom" in the nineties, affecting pure mathematicians as well as down-to-earth signal engineers and communication companies. Pete smelled right — once again!

In the following we summarize some resent results from our enterprise into an area where we find wavelets particularly useful: Kolmogorov's probabilistic cascade approach to intermittency in fully developed turbulence. In this approach anomalous fluctuations of the velocity as well as energy dissipation fields are modeled by hierarchically organized random breakups of vortices (or eddies). So far the intermittency phenomenon has not been derived from first principles (via the Navier-Stokes equations); hence, quite a variety of phenomenological models have been proposed in order to achieve a better understanding of the nature of such anomalous fluctuations. One train of thought emerged directly from Kolmogorov's legacy of the 60's and is represented by multiplicative cascade models, of which even the simplest member, the so-called p-model, describes some features of turbulence data remarkably well [1].

In general, such (for simplicity) one-dimensional models split some spatial domain, say bin, into two smaller subbins of equal length, whereas the energy density $\epsilon_k^{(j)}$ contained in the parent bin is divided randomly into two unequal parts $\epsilon_{2k}^{(j+1)} = (1+q)\epsilon_k^{(j)}$ and $\epsilon_{2k+1}^{(j+1)} = (1-q)\epsilon_k^{(j)}$. The splitting parameter q is drawn from a probability distribution $p(q)$, also called splitting function. This cascade prescription is then iterated for every cascade step j and every bin k, where all realizations of the weight q are drawn mutually independent.

After some J=8–12 cascade steps the model configurations have striking "similarity" to experimentally observed snapshots of the energy dissipation field in fully developed turbulent flows, both in atmospheric and hydrodynamic systems. Much effort has been invested into the analysis and quantification of this "similarity" in terms of (multi-)fractals (which is a fancy name for the scaling properties of moments $M_q(j) = \frac{1}{2^j}\sum_{k=0}^{2^j-1}(\epsilon_k^{(j)})^q$) and quite good agreement of the multifractal properties of such cascade models and experimental data has been reported repeatedly.

If we are to take such cascade models seriously, we need to clarify if this hierarchy of random-breakups is present in observed turbulence; this, however, cannot be unambiguously deduced from multifractal analyses alone. One requires a more 'local' analysis, i.e. one needs to study and compare *spatial* correlation functions in order to get further hints on the validity of Kolmogorov's initial ideas. The spatial correlation functions for a general class of hierarchical cascade processes (including the p-model) have been calculated in Refs. 2 and 3. The former are defined as the configuration average of a multiple product of energy densities, i.e. $\rho_{k_1,\ldots,k_q} = \langle \epsilon_{k_1}^{(J)} \cdot \ldots \cdot \epsilon_{k_q}^{(J)} \rangle$, and can be obtained from a generating function $Z[\lambda] = \langle \exp(i\sum_k \lambda_k \epsilon_k^{(J)}\rangle$ by taking derivatives with respect to the conjugate variables λ_k:

$$\rho_{k_1,\ldots,k_q} = \left\langle \epsilon_{k_1}^{(J)} \cdot \ldots \cdot \epsilon_{k_q}^{(J)} \right\rangle = \frac{1}{i^q} \left. \frac{\partial^q Z[\lambda]}{\partial \lambda_{k_1} \cdots \partial \lambda_{k_q}} \right|_{\vec{\lambda}=0} \quad ; \tag{1}$$

The generating function itself can be constructed from iterations of a backward or forward [4] evolution equation; here we state the former:

$$Z^{(j)}[\lambda^{(j)}] = \int \mathrm{d}q\, p(q)\; Z^{(j-1)}[(1+q)\lambda_L^{(j-1)}]\; Z^{(j-1)}[(1-q)\lambda_R^{(j-1)}] \quad , \tag{2}$$

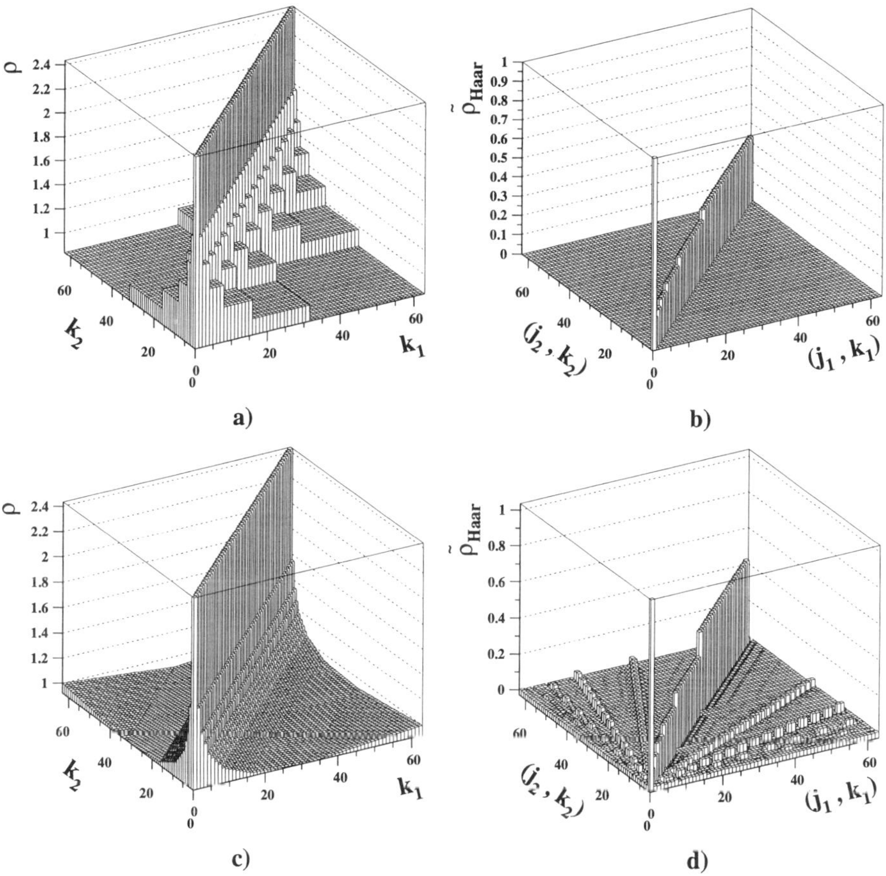

Figure 1: Two-bin correlation density $\rho_{k_1 k_2}$ for (a) a binary turbulent cascade model and (c) the same model made translational invariant by averaging over two adjacent independent configurations. (b) and (d) are the Haar-wavelet transformed correlation densities of (a) and (c) respectively.

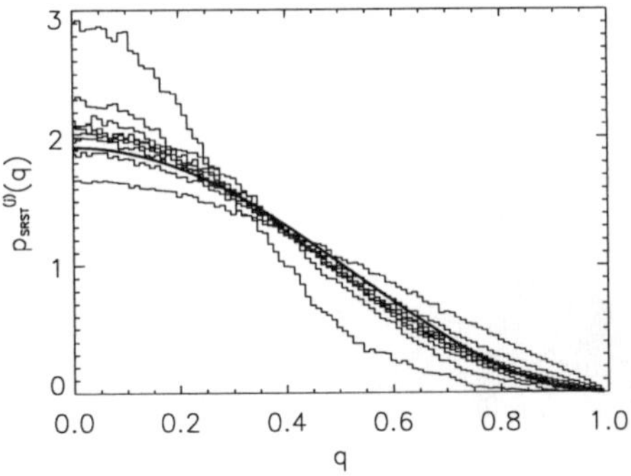

Figure 2: Scale dependent discretized probability distribution function $p^{(j)}(q)$ of the splitting weights obtained from the restoration scheme for translational invariance with two independent adjacent turbulent cascade model configurations. The solid curve corresponds to the experimentally obtained splitting function $p_{\exp}(q) = (1 - q^2)^{2.2}$ of Ref. [5].

where the indices L/R denote the left or right subbins, respectively. Insertion of (2) into (1) yields recursion relations for the spatial correlation functions.

Figure 1a shows the two-bin correlation density for a turbulent cascade model, where the experimentally deduced[5] splitting function $p(q) \sim (1-q^2)^{2.2}$ has been employed. The power-law rise towards the diagonal indicates the underlying selfsimilarity of the process. However, it is obvious that this correlation function is not invariant with respect to spatial translations, that is $\rho_{k_1,k_2} \neq \rho_{k_1+\Delta k,k_2+\Delta k}$. This is because of the binary hierarchical organization of the turbulent cascade model. However, correlation functions obtained from experimental observations (e.g. time series from anemometers) are translational invariant, because it is not possible to trigger on a cascading large eddy structure; hence, no knowledge on the beginning and ending of a cascade configuration is present.

In order to restore translational invariance in the spatial correlation functions, one needs to average over different cascade beginnings and endings just as the experimenters do in data taking; we simulate such an averaging process by joining two independent cascade configurations of equal size and shifting an artificial window of the same size randomly within the two independent configurations and determine the correlation functions based on the bins contained in the window. In this way, we also have no knowledge on the beginning and ending of a true cascade configuration. In Figure 1b the resulting two-bin correlation density $\langle \rho_{k_1 k_2} \rangle_t$ is depicted, which is now translational invariant.

It appears, however, that we have given up selfsimilarity in this approach to restore translational invariance. In other words: starting with a selfsimilar cascade process, the restoration of translational invariance will break the underlying selfsimilarity. This can be best elucidated once the scale-dependent splitting function $p^{(j)}(q)$ is extracted from configurations within the window randomly placed in two adjacent model configurations (see Fig. 2). The effective splitting functions at the beginning of the cascade (small j) is narrowly peaked around $q = 0$, whereas at the end of the cascade (large j) they become more flat. However, for a wide range of intermediate scales j, the effective splitting functions $p^{(j)}(q)$ are approximately equal to the original splitting function $p(q)$, which is shown in Fig. 2 as smooth solid line. As a consequence, the selfsimilarity is not severely broken by the scheme to restore translational invariance! Interestingly enough, such deviations of $p^{(j)}(q)$ from $p(q)$ are also in accordance with the experimental findings [5].

There is more to learn from the turbulent cascade models. So far the spatial correlation functions have been represented in the monoscale bin representation (see first row of Fig. 3). Obviously, for a hierarchically organized process a multiscale representation would be better suited. Such a representation is illustrated in the second row of Fig. 3 and is known as (Haar-) wavelet decomposition. The correlations between the corresponding (Haar-) wavelet amplitudes $\tilde{\epsilon}_{jk}$ can be obtained either by multidimensional wavelet transforms of the spatial correlations functions or by formulating the evolution equations directly in terms of $\tilde{\epsilon}_{jk}$. In second order, i.e. for $\tilde{\rho}_{(j_1k_1)(j_2k_2)} = \langle \tilde{\epsilon}_{j_1k_1} \tilde{\epsilon}_{j_2k_2} \rangle$ the results of the wavelet transform are depicted in Figs. 1b+d. Without restoration of translational invariance the Haar-wavelets represent the exact normal coordinates as they diagonalize the covariance matrix. With restoration of translational invariance the covariance matrix is still quasi diagonal. Hence, we conclude that wavelets represent good approximations to the true normal coordinates of the energy dissipation process in fully developed turbulence.

Given these results on the implications of the Haar-wavelet transformation for random multiplicative cascade models, one is immediately driven to the question: Are there more general cascade processes for which *any* given wavelet basis represents exactly the true normal coordinates? For example, the multiresolution analysis associated with Daubechies wavelets uses overlapping, but orthogonal low- and high-pass filters. Correspondingly, the more general cascade processes would have to reflect this spatially overlapping branching structure instead of the binary branching structure of the models used so far. In order to find the relevant cascade prescription and to determine their statistical properties, the approach presented in Refs. [2,3] needs to be modified: instead of a backward evolution equation for the multivariate generating func-

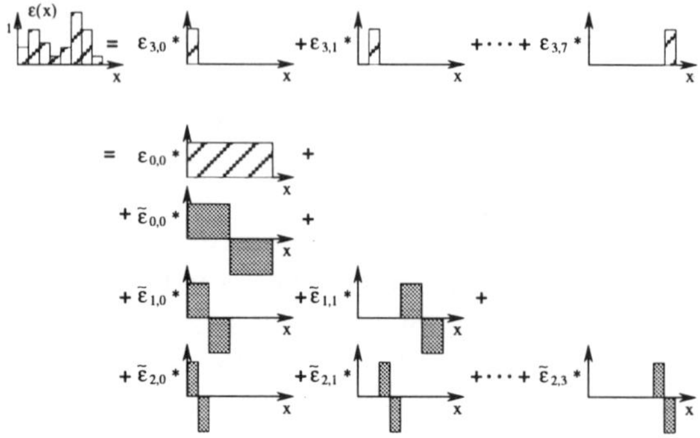

Figure 3: Monoscale bin representation (first row) and multiscale Haar-wavelet representation (second row) of one p-model configuration obtained after J=3 cascade steps.

tion of the spatial correlation densities a forward evolution equation is required to incorporate spatially overlapping branchings. In Ref. [4] we construct a specific example for which the Daubechies wavelets with 4 filter coefficients (D4) are the exact normal coordinates. This D4-wavelet cascade then has *overlapping* branches. Generalizations to other wavelet cascades (all with overlapping branches) are straightforward.

Fig. 4 depicts the evolution of one D4-wavelet cascade configuration (right) during the various stages. For comparison we also show a p-model configuration (left), for which in each branching the random weights q have been chosen identical to the ones in the illustrated D4-wavelet cascade configuration. Only for the first cascade step the D4-cascade configuration is identical to the p-model configuration. For all the further steps the two configurations deviate more and more because of the overlaps of the D4-cascade branchings. In the following we summarize the statistical properties of these D4-wavelet cascades.

The spatial second-order correlation function of the D4-wavelet cascade is shown in Fig. 5b. Compared to the one of the p-model in Fig. 5a, there are two principal differences. First, the diagonal elements are not all the same for the D4-wavelet cascade. The reason for this is that the moduli of the D4-filter coefficients c_0, c_1, c_2 and c_3 are not identical (as in the case for the Haar-wavelet, where $c_0 = c_1 = 1$). Second, we used periodic continuation of the D4-wavelet at the endpoints of the "observation" interval, which causes the second-order correlation density to increase for large bin-distances. Apart

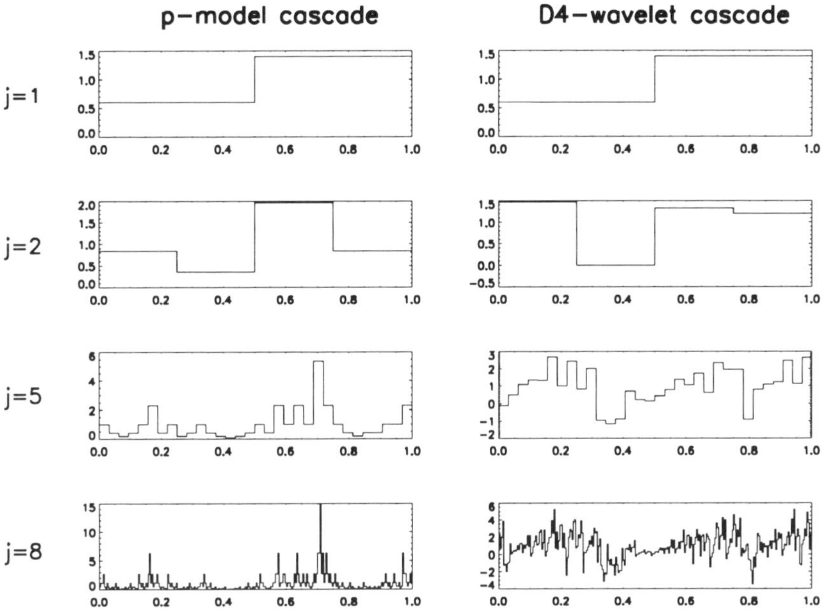

Figure 4: Evolution of one p-model configuration (left column) and one D4-wavelet cascade configuration (right column). The random multipliers $q_k^{(j)}$ have been identically chosen for the two cascade evolutions.

from these (technical) differences, it is obvious that the overlapping of branches leads to an "almost" translationally invariant correlation function. In this type of model the energy is not only transfered to the two immediate children of a parent bin, but rather "smeared out" over four adjacent bins proportional to the four weights c_i of the D4-scaling function filter.

Motivated from the p-model correlation results we switch again to a multiscale representation. Fig. 5d illustrates the Haar-wavelet correlation densities $\tilde{\rho}^{Haar}_{k_1,k_2}$ of the D4-wavelet cascade; the correlation matrix has become quasi-diagonal where off-diagonal elements arise in bands and are suppressed with respect to the diagonal elements. The diagonal elements show an approximate power-law in terms of their scale index j; this is even more obvious in Ref. 4 where we also discuss wavelet-moments constructed from the wavelet correlation densities.

The quasi-diagonality of the Haar-wavelet transformed correlation matrix indicates that the employed building blocks (Haar-wavelets) still do not represent an optimal choice to describe the various branchings occuring in the

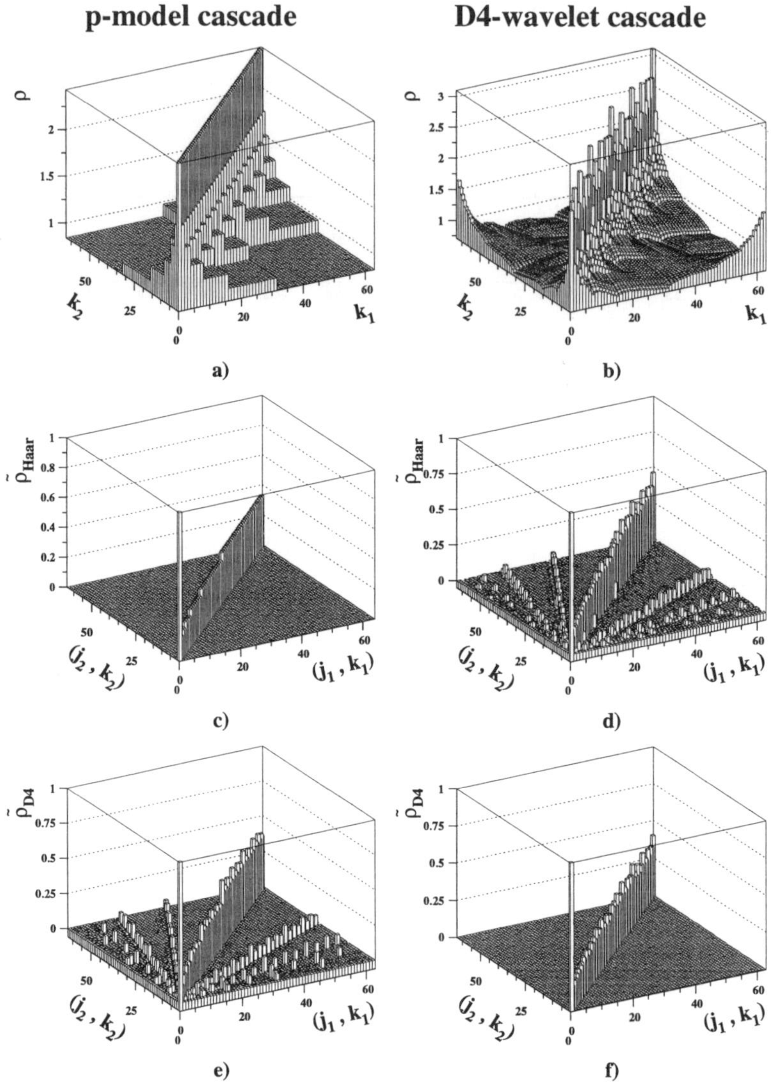

Figure 5: (a) Spatial two-bin correlation density of the p-model cascade after J=6 cascade steps, (c) its Haar-wavelet and (e) D4-wavelet transform. (b), (d) and (f) are the same as (a), (c) and (e), but for the D4-wavelet cascade.

D4-wavelet cascade. However, the D4-wavelet transformed correlation density turns out to be exactly diagonal; see Fig. 5f. If we would not know from it's construction, we would conclude that this cascade represents a selfsimilar process with the D4-wavelet as its true building blocks (hence the name D4-*wavelet cascade*)

Seen in a more general context, these results suggest the following strategy for a solid analysis of turbulence: in order to clarify the validity of Kolmogorov's picture, one needs to identify the true normal coordinates (building blocks) of the process first. The latter may be found by searching for a wavelet-like basis which best diagonalizes the correlation functions of second order and test if higher orders are consistently represented. Only then the true scale-dependence can be analyzed and the hierarchical breakup of localized energy clumps (eddies) be verified.

In the future this last claim has to be further supported from data and other empirical models. Furthermore, spatio-temporal correlations have to be studied; this could set the stage for a direct derivation of the turbulent cascade models from a Navier-Stokes equation with wavelet techniques.

As a last remark, we would like to mention that Pete's far-seeing advice has led to many further fruitful applications of wavelet-techniques beyond models for fully developed turbulence; Pete's Tucson colleagues are now using wavelet techniques to study multiparticle and jet dynamics [3], disoriented chiral condensates [6], disordered solid-state systems [7], quantum algebras [8] and last-not-least the large scale structure of the universe [9]. More is sure to come. *Thank you Pete!*

Acknowledgement: Work supported by DFG, GSI and BMBF. P.L. is indebted to the Österreichische Akademie der Wissenschaften for its support with an APART fellowship.

1. C. Meneveau and K.R. Sreenivasan, J. Fluid Mech. 224 (1991) 429.
2. M. Greiner, P. Lipa and P. Carruthers, Phys. Rev. E 51 (1995) 1948.
3. M. Greiner, J. Giesemann, P. Lipa and P. Carruthers, Z. Phys. C 69 (1996) 305.
4. J. Giesemann, M. Greiner and P. Lipa, preprint HEPHY-PUB 666-97.
5. K.R. Sreenivasan and G. Stolovitzky, J. Stat. Phys. 78 (1995) 311.
6. Z. Huang, I. Sarcevic, R. Thews and X.-N. Wang, Phys. Rev. D 54 (1996) 750.
7. J. Kantelhardt, M. Greiner and E. Roman, Physica A 220 (1995) 219; E. Roman, J. Kantelhardt and M. Greiner, Europhys. Lett. 35 (1996) 641.
8. A. Ludu and M. Greiner, *ICTP-Internal Report* IC/95/214.
9. J. Pando and L.Z. Fang, Astrophys. J. 459 (1996) 1; J. Pando, P. Lipa, M. Greiner and L.Z. Fang, preprint AZPH-TH/97-03.

CORRELATIONS AND FLUCTUATIONS IN HIGH-ENERGY COLLISIONS

W. KITTEL
HEFIN, University of Nijmegen,
Toernooiveld 1, 6525 ED Nijmegen, The Netherlands

In addition to its importance in describing high-energy processes themselves, the dynamics of multiparticle production is part of the general field of non-linear phenomena and complex systems. Multiparticle dynamics is one of the rare fields of physics where *higher*-order correlations are directly accessible in their full multi-dimensional characteristics under well-controlled experimental conditions. Multiparticle dynamics, therefore, is an ideal testing ground for the development of advanced statistical methods.
Higher-order correlations have, indeed, been observed as particle-density fluctuations. Approximate scaling with finer resolution provides evidence for a self-similar correlation effect. Quantum-Chromodynamics branching is a good candidate for a dynamical explanation of these correlations in e^+e^- collisions at CERN/LEP and, as expected, also of those in pp collisions at future CERN/LHC energies. However, other sources such as identical-particle Bose-Einstein interference effects also contribute.
A particular question at the moment is the smooth transition from the QCD branching domain (gluon interference before hadronization) to the Bose-Einstein domain (identical-pion interference after hadronization). Both mechanisms have clearly been observed in e^+e^- collisions at CERN/LEP energies. The large amount of high-resolution data being collected at LEP will allow the study of the *genuine* (i.e. non-trivial) higher-order correlations in both domains.

1 Introduction

Recent years have witnessed a remarkable amount of experimental and theoretical activity in search of scale-invariance and fractality in multihadron production processes. In addition to being an important part of high energy physics itself, the dynamics of multi-particle production in collisions of elementary particles at high energies (multiparticle dynamics) is part of the general field of non-linear phenomena and complex systems. Studies of classical and quantum chaos, non-equilibrium dissipative processes, random media, growth phenomena and many more processes have all contributed in revealing the pervasive importance of self-similarity, power-laws, and fractals in nature. Research in these fields is still in full swing and continues to uncover intriguingly simple and often surprisingly universal behavior in complex, non-linear systems.

While considerable experience already exists in many fields for the study of two-component correlation, it is often in *higher*-order (i.e. multi-component) correlations that the most interesting properties manifest themselves, in the simultaneous interplay of a large number of components. The special significance of multiparticle dynamics for the development of advanced statistical methods lies in the fact that it is one of the rare fields of physics where higher-order correlations are directly accessible in their full multi-dimensional characteristics under well controlled experimental conditions.

Higher-order correlations have recently been observed as particle-density fluctuations in cosmic ray, nucleus-nucleus, hadron-hadron, e^+e^- and lepton-hadron experiments. To study these fluctuations in detail, normalized correlation integrals are being analyzed in phase-space domains of ever decreasing size. Approximate scaling with decreasing domain size is now observed in all types of collision, giving evidence for a correlation effect self-similar over a large range of the resolution (called intermittency, in analogy to a statistically similar problem in spatio-temporal turbulence).

Parton branching of Quantum-Chromodynamics predicts the type of correlations observed in e^+e^- collisions at CERN/LEP and, as expected, also in pp collisions at CERN/LHC energies. However, other sources such as Bose-Einstein interference of identical particles also contribute. Fast development of the applied technology has taken place over the last few years, in particular in the extension of originally one-dimensional to full three-dimensional phase space analysis.

2 Methodology

2.1 Particle Densities

A collision between particles a and b is assumed to yield exactly n particles in a sub-volume Ω of total phase space Ω_{all}. The single symbol y represents the kinematical variables needed to specify the position of each particle in this space (for example, y can be the full four-momentum of a particle and Ω a cell in invariant phase space or simply the c.m. rapidity [a] of a particle and Ω an interval of length δy). The distribution of points in Ω can then be characterized by continuous probability densities $P_n(y_1, \ldots, y_n)$; $n = 1, 2, \ldots$. For

[a] Rapidity y is defined as $y = \frac{1}{2}\ln[(E + p_{\text{L}})/(E - p_{\text{L}})]$, with E the energy and p_{L} the longitudinal component of momentum vector $\mathbf{p}$ along a given direction (beam-particles, jet-axis, etc.); pseudo-rapidity is defined as $\eta = \frac{1}{2}\ln[(p + p_{\text{L}})/(p - p_{\text{L}})]$.

simplicity, we assume all final-state particles to be of the same type. In this case, the **exclusive** distributions $P_n(y_1,\ldots,y_n)$ can be taken as fully symmetric in $y_1,\ldots,y_n$; they describe the distribution in Ω when the multiplicity is exactly n.

The corresponding **inclusive** distributions are given for $n = 1,2,\ldots$ by:

$$\begin{aligned}\rho_n(y_1,\ldots,y_n) &= P_n(y_1,\ldots,y_n) \\ &+ \sum_{m=1}^{\infty} \frac{1}{m!} \int_\Omega P_{n+m}(y_1,\ldots,y_n,y'_1,\ldots,y'_m) \prod_{i=1}^{m} \mathrm{d}y'_i \,. \end{aligned} \quad (1)$$

The inverse formula is

$$\begin{aligned}P_n(y_1,\ldots,y_n) &= \rho_n(y_1,\ldots,y_n) \\ &+ \sum_{m=1}^{\infty} (-1)^m \frac{1}{m!} \int_\Omega \rho_{n+m}(y_1,\ldots,y_n,y'_1,\ldots,y'_m) \prod_{i=1}^{m} \mathrm{d}y'_i \,. \end{aligned} \quad (2)$$

Here, $\rho_n(y_1,\ldots,y_n)$ is the probability density for n points to be at $y_1,\ldots,y_n$, irrespective of the presence and location of any further points. Integration over an interval Ω in y yields

$$\begin{aligned} &\int_\Omega \rho_1(y)dy = \langle n \rangle \,, \\ &\int_\Omega \int_\Omega \rho_2(y_1,y_2)dy_1dy_2 = \langle n(n-1) \rangle \,, \\ &\int_\Omega dy_1 \ldots \int_\Omega dy_q \rho_q(y_1,\ldots,y_q) = \langle n(n-1)\ldots(n-q+1) \rangle \,, \end{aligned} \quad (3)$$

where the angular brackets imply the average over the event ensemble.

2.2 Cumulant Correlation Functions

Besides the interparticle *correlations* we are looking for, the inclusive q-particle densities $\rho_q(y_1,\ldots,y_q)$ in general contain "trivial" contributions from lower-order densities. It is, therefore, advantageous to consider a new sequence of functions $C_q(y_1,\ldots,y_q)$ as those statistical quantities which vanish whenever one of their arguments becomes statistically independent of the others. Deviations of these functions from zero shall be addressed as *genuine* correlations.

The quantities with the desired properties are the correlation functions - also called (factorial) cumulant functions - or, in integrated form, Thiele's semi-invariants.[1] A formal proof of this property was given by Kubo.[2] The cumulant correlation functions are defined as in the cluster expansion familiar from statistical mechanics via the sequence: [3,4,5]

$$\rho_1(1) = C_1(1), \tag{4}$$
$$\rho_2(1,2) = C_1(1)C_1(2) + C_2(1,2), \tag{5}$$
$$\rho_3(1,2,3) = C_1(1)C_1(2)C_1(3) + C_1(1)C_2(2,3) + C_1(2)C_2(1,3) + + C_1(3)C_2(1,2) + C_3(1,2,3); \tag{6}$$

and, in general, by

$$\rho_m(1,\ldots,m) = \sum_{\{l_i\}_m} \sum_{\text{perm.}} \underbrace{[C_1()\cdots C_1()]}_{l_1\,\text{factors}} \underbrace{[C_2(,)\cdots C_2(,)]}_{l_2\,\text{factors}} \cdots \cdots \underbrace{[C_m(,\ldots,)\cdots C_m(,\ldots,)]}_{l_m\,\text{factors}}. \tag{7}$$

Here, l_i is either zero or a positive integer and the sets of integers $\{l_i\}_m$ satisfy the condition

$$\sum_{i=1}^{m} i\, l_i = m. \tag{8}$$

The arguments in the C_i functions are to be filled by the m possible momenta in any order. The sum over permutations is a sum over all distinct ways of filling these arguments. For any given factor product there are precisely [4]

$$\frac{m!}{[(1!)^{l_1}(2!)^{l_2}\cdots(m!)^{l_m}]\, l_1! l_2! \cdots l_m!} \tag{9}$$

terms.

The relations (7) may be inverted with the result:

$$C_2(1,2) = \rho_2(1,2) - \rho_1(1)\rho_1(2)\,,$$
$$C_3(1,2,3) = \rho_3(1,2,3) - \sum_{(3)} \rho_1(1)\rho_2(2,3) + 2\rho_1(1)\rho_1(2)\rho_1(3)\,,$$
$$C_4(1,2,3,4) = \rho_4(1,2,3,4) - \sum_{(4)} \rho_1(1)\rho_3(1,2,3) - \sum_{(3)} \rho_2(1,2)\rho_2(3,4) + 2\sum_{(6)} \rho_1(1)\rho_1(2)\rho_2(3,4) - 6\rho_1(1)\rho_1(2)\rho_1(3)\rho_1(4). \tag{10}$$

In the above relations we have abbreviated $C_q(y_1,\ldots,y_q)$ to $C_q(1,2,\ldots,q)$; the summations indicate that all possible permutations must be taken (the number under the summation sign indicates the number of terms). Expressions for higher orders can be derived from the related formulae given in [6].

It is often convenient to divide the functions ρ_q and C_q by the product of one-particle densities, which leads to the definition of the normalized inclusive densities and correlations:

$$r_q(y_1,\ldots,y_q) = \rho_q(y_1,\ldots,y_q)/\rho_1(y_1)\ldots\rho_1(y_q), \quad (11)$$

$$K_q(y_1,\ldots,y_q) = C_q(y_1,\ldots,y_q)/\rho_1(y_1)\ldots\rho_1(y_q). \quad (12)$$

In terms of these functions, correlations have been studied extensively for $q = 2$. Results also exist for $q = 3$, but usually the statistics (i.e. number of events available for analysis) are too small to isolate genuine correlations. To be able to do that for $q \geq 3$, one must apply moments defined via the integrals in Eq. 3, but in limited phase-space cells.

2.3 Cell-Averaged Factorial Moments and Cumulants

In practical work, with limited statistics, it is almost always necessary to perform averages over more than a single phase-space cell. Let Ω_m be such a cell (e.g. a single rapidity interval of size δy) and divide the phase-space volume into M non-overlapping cells Ω_m of size $\delta\Omega$ independent of m. Let n_m be the number of particles in cell Ω_m. Different cell-averaged moments may be considered, depending on the type of averaging.

Normalized cell-averaged factorial moments [7] are defined as

$$F_q(\delta y) \equiv \frac{1}{M}\sum_{m=1}^{M}\frac{\langle n_m(n_m-1)\ldots(n_m-q+1)\rangle}{\langle n_m\rangle^q} \quad (13)$$

$$\equiv \frac{1}{M}\sum_{m=1}^{M}\frac{\int_{\delta y}\rho_q(y_1,\ldots,y_q)\prod_{i=1}^{q}dy_i}{\left(\int_{\delta y}\rho(y)dy\right)^q} \quad (14)$$

$$= \frac{1}{M(\delta y)^q}\sum_{m=1}^{M}\int_{\delta y}\frac{\rho_q(y_1,\ldots,y_q)\prod_{i=1}^{q}dy_i}{(\bar{\rho}_m)^q}. \quad (15)$$

The full rapidity interval ΔY is divided into M equal bins: $\Delta Y = M\delta y$; each y_i is within the δy-range and $\langle n_m\rangle \equiv \bar{\rho}_m\delta y \equiv \int_{\delta y}\rho_1(y)dy$. An example for $q = 2$ is given in Fig. 1a.

Likewise, cell-averaged normalized factorial cumulant moments may be defined as

$$K_q(\delta y) = \frac{1}{M(\delta y)^q}\sum_{m=1}^{M}\int_{\delta y}\frac{C_q(y_1,\ldots,y_q)\prod_{i=1}^{q}dy_i}{(\bar{\rho}_m)^q} \quad . \quad (16)$$

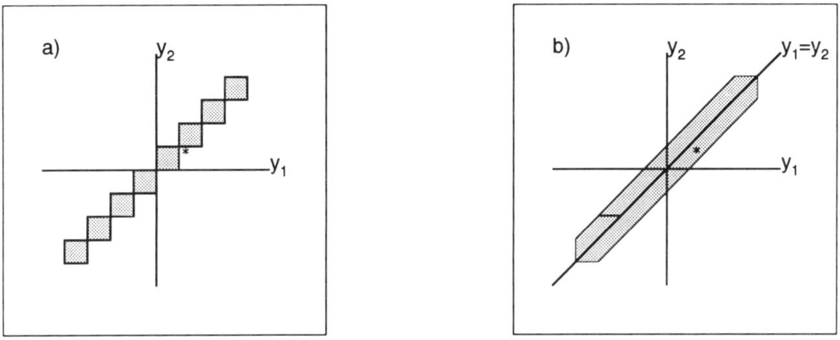

Figure 1: Integration domains for a) the second-order factorial moment and b) the second-order correlation integral. The asterisk in a) indicates the position of a particle pair with $|y_1 - y_2| < \delta y$ that is excluded from the F_2 calculation due to the binning. In b) the asterisk is included.

They are related to the factorial moments by [b]

$$\begin{aligned} F_2 &= 1 + K_2 \quad , \\ F_3 &= 1 + 3K_2 + K_3 \quad , \\ F_4 &= 1 + 6K_2 + 3\overline{K_2^2} + 4K_3 + K_4 \, . \end{aligned} \tag{17}$$

In F_4 and higher-order moments, "bar averages" appear defined as $\overline{AB} \equiv \sum_m A_m B_m / M$.

To detect dynamical fluctuations in the density of particles produced in a high-energy collision, a way must be devised to eliminate, or to reduce as much as possible, the statistical fluctuations (noise) due to the finiteness of the number of particles in the counting cell(s). This requirement can to a large extent be satisfied by studying factorial moments and forms the basis of the factorial moment technique, known in optics, but rediscovered for multi-hadron physics in [7]. This crucial property does not apply to, e.g., ordinary moments $\langle n^q \rangle / \langle n \rangle^q$.

The property of Poisson-noise suppression has made measurement of factorial moments a standard technique, e.g. in quantum optics, to study the statistical properties of arbitrary electromagnetic fields from photon-counting distributions. Their utility was first explicitly recognized, for the single time-interval case, in [9] and later generalized to the multivariate case in [10].

[b] The higher-order relations can be found in [8].

2.4 Density and Correlation Integrals

A fruitful recent development in the study of density fluctuations is the correlation strip-integral method.[11] By means of integrals of the inclusive density over a strip domain of Fig. 1b, rather than a sum of the box domains of Fig. 1a, one not only avoids unwanted side-effects such as splitting of density spikes, but also drastically increases the integration volume (and therefore the statistical significance) at a given resolution.

Let us consider first the factorial moments F_q defined according to Eq. 14. As shown in Fig. 1a for $q = 2$, the integration domain $\Omega_B = \sum_{m=1}^{M} \Omega_m$ consists of M q-dimensional boxes Ω_m of edge length δy. A point in the m-th box corresponds to a pair (y_1, y_2) of distance $|y_1 - y_2| < \delta y$ and both particles in the same bin m. Points with $|y_1 - y_2| < \delta y$ which happen *not* to lie in the same, but in adjacent, bins (e.g., the asterisk in Fig. 1a) are left out. The statistics can be approximately doubled by a change of the integration volume Ω_B to the strip-domain of Fig. 1b. For $q > 2$, the increase of integration volume (and reduction of squared statistical error) is in fact roughly proportional to the order of the correlation. The gain is even larger when working in two- or three-dimensional phase-space variables.

In terms of the strips (or hyper-tubes for $q > 2$), the density integrals become

$$F_q^{\mathrm{S}}(\delta y) \equiv \frac{\int_{\Omega_s} \rho_q(y_1 \dots y_q) \prod_i \mathrm{d}y_i}{\int_{\Omega_s} \rho_1(y_1) \dots \rho_1(y_q) \prod_i \mathrm{d}y_i} \quad . \tag{18}$$

These integrals can be evaluated directly from the data after selection of a proper distance measure ($|y_i - y_j|$, $[(y_i - y_j)^2 + (\phi_i - \phi_j)^2]^{1/2}$, or better yet, the four-momentum difference $Q_{ij}^2 = -(p_i - p_j)^2$) and after definition of a proper multiparticle topology (GHP integral,[11] snake integral,[12] star integral [13]). Similarly, *correlation* integrals can be defined by replacing the density ρ_q in Eq. 18 by the correlation function C_q.

The numerator of the factorial moments F_q can be determined by counting, for each event, the number of q-tuples that have a pairwise Q_{ij}^2 smaller than a given value Q^2 and then averaging over all events. Using the Heaviside unit-step function Θ, this can be mathematically expressed as

$$F_q^{\mathrm{S}}(Q^2) = \frac{1}{\mathrm{norm}} \left\langle q! \sum_{i_1 < \dots < i_q} \prod_{\substack{all\ pairs \\ k_1, k_2}} \Theta(Q^2 - Q_{i_{k_1}, i_{k_2}}^2) \right\rangle \quad , \tag{19}$$

where the factor $q!$ takes into account the number of permutations within a q-tuple.

The normalization is obtained from "mixed" events constructed by random selection of tracks from different events in a track pool. The multiplicity of a mixed event is taken to be a Poissonian random variable, thereby ensuring that no extra correlations are introduced. A correction factor is applied for the difference in average multiplicity of the Poissonian and the experimental distribution. The mixed events are treated in the same way as real events.

2.5 Power-Law Scaling

The technique proposed in [7] consists of measuring the dependence of the normalized factorial moments (or correlation integrals) $F_q(\delta y)$ as a function of the resolution δy. For definiteness, δy is supposed to be an interval in rapidity, but the method generalizes to arbitrary phase-space dimension, as occurs with the use of Q^2.

As pointed out above, the scaled factorial moments enjoy the property of "noise-suppression". High-order moments further act as a filter and resolve the large n_m tail of the multiplicity distribution. They are thus particularly sensitive to large density fluctuations at the various scales δy used in the analysis.

As proven in [7], a "smooth" (rapidity) distribution, which does not show any fluctuations except for the statistical ones, has the property of $F_q(\delta y)$ being independent of the resolution δy in the limit $\delta y \to 0$. On the other hand, if dynamical fluctuations exist and P_ρ is "intermittent", the F_q obey the power law

$$F_q(\delta y) \propto (\delta y)^{-\phi_q} \ , \quad (\delta y \to 0). \tag{20}$$

The powers ϕ_q (slopes in a double-log plot) are related [14] to the anomalous dimensions

$$d_q = \phi_q/(q-1) \quad , \tag{21}$$

a measure for the deviation from an integer dimension. Equation 20 is a scaling law since the ratio of the factorial moments at resolutions L and ℓ

$$R = F_q(\ell)/F_q(L) = (L/\ell)^{\phi_q} \tag{22}$$

only depends on the ratio L/ℓ, but not on L and ℓ, themselves.

As pointed out above, the experimental study of correlations is already difficult for three particles. The close connection between correlations and factorial moments offers the possibility of measuring higher-order correlations with the factorial moment technique at smaller distances than was previously feasible. Via Eq. 21, the method further relates possible scaling behavior of such correlations to the physics of fractal objects.

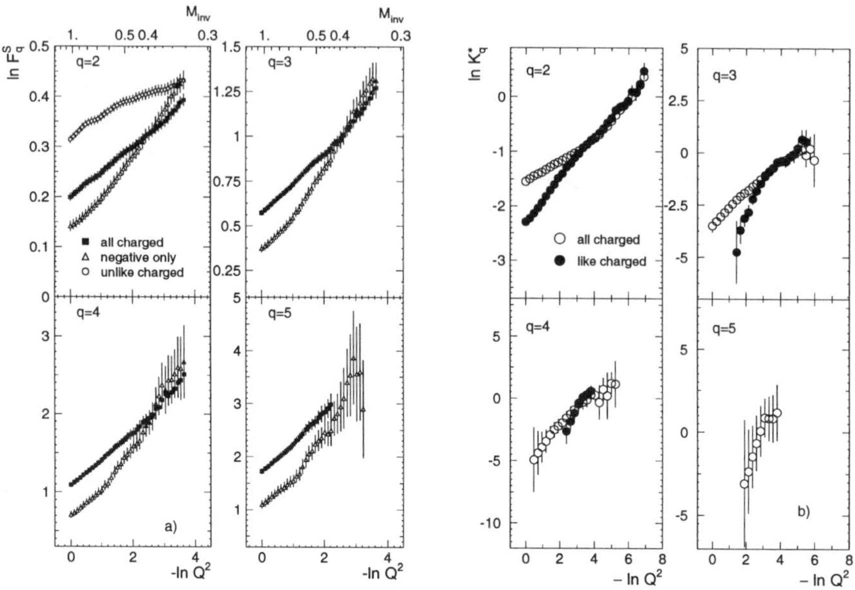

Figure 2: Correlation-integral method applied to NA22 hadron-hadron data [17,19] in terms of a) $\ln F_q^S$ and b) $\ln K_q^*$ as a function of $-\ln Q^2$.

One further has to stress the advantages of factorial cumulants compared to factorial moments, since the former measure *genuine* correlation patterns, whereas the latter contain additional large combinatorial terms which mask the underlying dynamical correlations.

The definition of "intermittency" given in (20), has its origin in other disciplines.[c] It rests on a loose parallel between the high non-uniformity of the distribution of energy dissipation, for example, in turbulent intermittency and the occurrence of large "spikes" in hadronic multiparticle final states. In the following we use the term "intermittency" in a weaker sense, meaning the rise of factorial moments with increasing resolution, not necessarily according to a strict power law.

3 The State of the Art

The suggestion that normalized factorial moments of particle distributions might show power-law behavior has spurred a vigorous experimental search

[c] For a masterly exposé of this subject see [15].

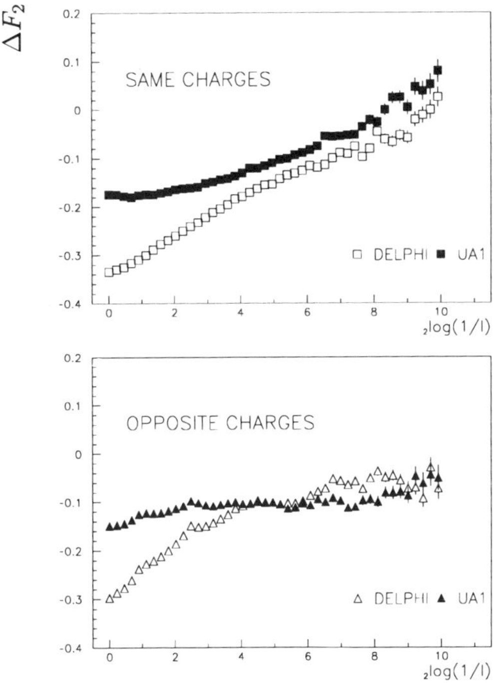

Figure 3: Comparison of density integrals for $q = 2$ in their differential form ΔF_2 (in intervals $Q^2, Q^2 + dQ^2$) as a function of $_2\log(1/Q^2)$ for e^+e^- (DELPHI, preliminary) and hadron-hadron collisions (UA1).[20]

for (more or less) linear dependence of $\ln F_q$ on $-\ln \delta y$. A full review of the present situation is given in [16].

As an example, we give Fig. 2a where hadron-hadron data [17] on F_q^S are plotted as a function of $-\ln Q^2$, with all two-particle combinations in an n-tuple having $Q_{ij}^2 < Q^2$. The following observations can be made:

i) the moments show a steep rise with decreasing Q^2;

ii) negatives are much steeper than all-charged,

iii) F_2^S is flatter for $(+-)$ than for all-charged or like-charged combinations.

The last two observations directly demonstrate the large influence of *identical* particle correlations on the factorial moments and their scaling behavior. These results agree very well with results from the UA1 collaboration.[18] In Fig. 2b it is, furthermore, shown [19] in terms of the $K_q^*(Q^2)$ (using the star topology) that genuine correlations exist and increase with improving resolution at least up to order $q = 5$ in hadron-hadron collisions. Again, like-charged particle combinations increase faster than all-charged ones.

Of particular interest is a comparison of hadron-hadron to e^+e^- results in terms of same and opposite charges of the particles involved. This comparison has been done for UA1 and DELPHI data in [20] and is shown in Fig. 3 for $q = 2$ (in fact, in this figure a differential form ΔF_2 of Eq. 19 is presented). An important difference between UA1 and DELPHI can be observed in a comparison of the two sub-figures:

For relatively large $Q^2 (> 0.03 \text{ GeV}^2)$, where Bose-Einstein effects do not play a major role, the e^+e^- data increase much faster with increasing $-_2 \log Q^2$ than the hadron-hadron results. For e^+e^-, the increase in this Q^2 region is very similar for same and for opposite-sign charges.

At small Q^2, however, the e^+e^- results approach the hadron-hadron results.

The authors conclude that for e^+e^- annihilation at LEP at least two processes are responsible for the power-law behavior: Bose-Einstein correlation following the evolution of jets. In hadron-hadron collisions at present collider energies, the Bose-Einstein effects seem mostly relevant.

The exact functional form of ΔF_2 or ΔK_2 is derived from the data of UA1 [18,21] and NA22,[17] again in its differential form, in Fig. 4. Clearly, the low-Q data favor a power law in Q over a (non-scaling) exponential, double-exponential or Gaussian law.

In Fig. 5, the NA22 results for two-, three- and four-particle Bose-Einstein correlations (equivalent to $\Delta F_2(Q^2)$) are compared to a multi-Gaussian parametrization.[24] In the conventional linear presentation of Fig. 5a the fits look reasonable. If the same data and same fits are presented in a log-log scale, however, the power-law deviation from the multi-Gaussians starts to become visible (Fig. 5b).

If the observed effect is real, it supports a view recently developed in [24]. In that paper, intermittency is explained from Bose-Einstein correlations between (like-sign) pions. As such, Bose-Einstein correlations from a static source are not power-behaved. A power law is obtained if i) the size of the interaction region is allowed to fluctuate, and/or ii) the interaction region itself is assumed to be a self-similar object extending over a large volume. Condition ii) would be realized if parton avalanches were to arrange themselves into self-organized critical states.[25] Though quite speculative at this moment, it is an interesting new idea with possibly far-reaching implications. We should mention also that in such a scheme intermittency is viewed as a final-state interaction effect and is therefore not troubled by hadronization effects.

In perturbative QCD, on the other hand, the fractal structure of jets fol-

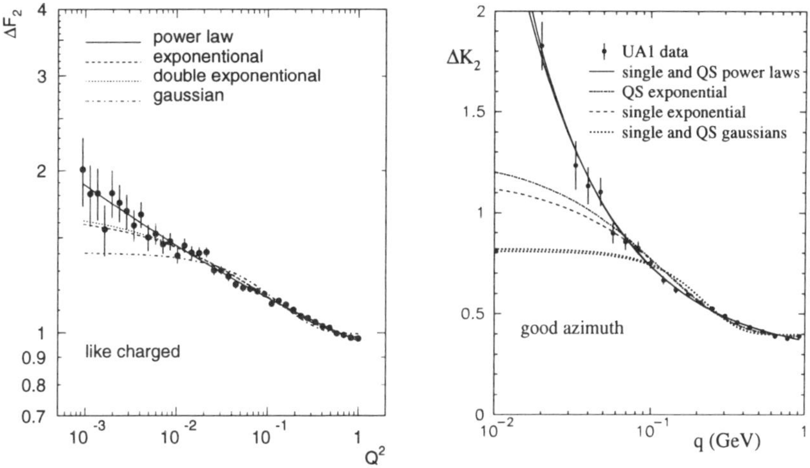

Figure 4: Density integral ΔF_2 as a function of $\log Q^2$ for NA22 [17], correlation integral ΔK_2 as a function of $\log Q$ for like-charged pairs in UA1 (preliminary) [21] compared to power-law, exponential, double-exponential and Gaussian fits, as indicated.

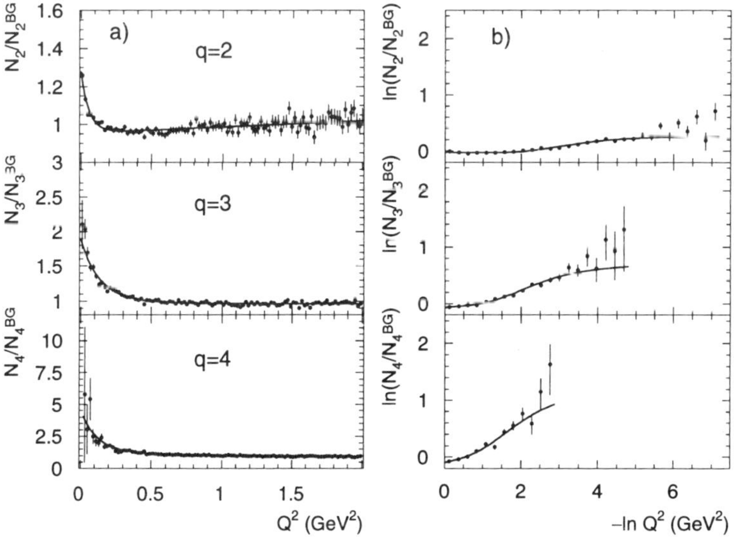

Figure 5: The normalized two-, three- and four-particle inclusive densities (equivalent to ΔF_q) as a function of Q^2 (left) and $-\ln Q^2$ (right).[22] Curves show the multi-Gaussian fits according to [23].

lows, in principle, from parton branching and the intermittency indices ϕ_q are directly related to the anomalous multiplicity dimension $\gamma_0 = (6\alpha_s/\pi)^{1/2}$ [26,27,28,29] and, therefore, to the running coupling constant α_s via the simple relation,

$$\phi_q = (q-1)D - \frac{q^2-1}{q}\gamma_0 \ ,$$

where D is the usual topological dimension of the analysis. In the same theoretical context, it has been argued [27,28,29] that the opening angle between particles is a suitable and sensitive variable to analyze and well suited for these first analytical QCD calculations of higher-order correlations. It is, of course, closely related to Q^2.

A first analytical QCD calculation [27] is based on the double-log approximation (DLA) with angular ordering [30] (for a recent experimental study of angular ordering see [31]) and on local parton-hadron duality.[32] A preliminary comparison with DELPHI data[33] gives encouraging results, but shows that the situation is far from trivial, since finite-energy effects, four-momentum cut-offs, resonance decays etc. still dominate at LEP energies.

4 Summary

Multiparticle production in high-energy collisions is an ideal field to study genuine higher-order correlations. Methods also used in other fields are being tested and extended here for general application. Indications for genuine, approximately self-similar higher-order correlations are indeed found in hadron-hadron collisions, but still need to be establised in their genuine and self-similar character in e^+e^- collisions at high energies. At large four-momentum distance Q^2, they are not only expected to be an inherent property of perturbative QCD, but are directly related to the anomalous multiplicity dimension and, therefore, to the running coupling constant α_s. At small Q^2, the QCD effects are complemented by Bose-Einstein interference of identical mesons carrying information on the unknown space-time development of particle production during the collision. The interplay between these two mechanisms, particularly important for an understanding of the process of hadronization, is completely unknown at the moment.

References

1. T.N. Thiele in *The Theory of Observation*, *Ann. Math. Stat.* **2** 165 (1931).
2. R. Kubo, *J. Phys. Soc. Japan* **17**, 1100 (1962).

3. B. Kahn, G.E. Uhlenbeck, *Physica* **5**, 399 (1938).
4. K. Huang in *Statistical Mechanics*, John Wiley and Sons, 1963.
5. A.H. Mueller, *Physics Review* D **4**, 150 (1971).
6. M.G. Kendall and A. Stuart in *The Advanced Theory of Statistics*, Vol. 1, C. Griffin and Co., London 1969.
7. A. Białas and R. Peschanski, *Nucl. Phys.* B **273**, 703 (1986); ibid. B **308**, 857 (1988).
8. P. Carruthers, H.C. Eggers and I. Sarcevic, *Phys. Lett.* B **254**, 258 (1991).
9. G. Bédard, *Proc. Phys. Soc.* **90**, 131 (1967).
10. D. Cantrell, *Phys. Rev.* A **1**, 672 (1970).
11. H.G.E. Hentschel and I. Procaccia, *Physica* D **8**, 435 (1983); P. Grassberger, *Phys. Lett.* A **97**, 227 (1983); I.M. Dremin, *Mod. Phys. Lett.* A **13**, 1333 (1988); P. Lipa, P. Carruthers, H.C. Eggers and B. Buschbeck, *Phys. Lett.* B **285**, 300 (1992).
12. P. Carruthers and I. Sarcevic, *Phys. Rev. Lett.* **63**, 1562 (1989).
13. H.C. Eggers *et al.*, *Phys. Rev.* D **48**, 2040 (1993).
14. P. Lipa and B. Buschbeck, *Phys. Lett.* B **223**, 465 (1989); R. Hwa, *Phys. Rev.* D **41**, 1456 (1990).
15. Ya.B. Zeldovich, A.A. Ruzmaikin, D.D. Sokoloff in *The Almighty Chance*, World Scientific Lecture Notes in Physics, Vol. 20 (World Scientific Singapore, 1990).
16. E.A. De Wolf, I.M. Dremin, W. Kittel, *Phys. Reports* **270** (1996) 1.
17. N. Agababyan *et al.*, NA22 Coll., *Z. Phys.* C **59**, 405 (1993).
18. N. Neumeister *et al.*, UA1 Coll., *Z. Phys.* C **60**, 633 (1993).
19. N.M. Agababyan *et al.*, NA22 Coll., *Z. Phys.* B **332**, 458 (1994).
20. F. Mandl and B. Buschbeck in *Proc. Cracow Workshop on Multiparticle Production, Krakow*, eds. A. Białas *et al.* (World Scientific, Singapore, 1994) p.1.
21. H. Eggers in *Proc. 7th Int. Workshop on Multiparticle Production Correlations and Fluctuations, Nijmegen*, eds. R. Hwa et al. (World Scientific, Singapore, to be publ.).
22. N.M. Agababyan *et al.*, NA22 Coll., *Z. Phys.* C **68**, 229 (1995) and W. Kittel in *Proc. XXVIth Int. Symp. on Multiparticle Dynamics, Faro*, eds. J. Dias de Deus et al., (World Scientific, Singapore, to be publ.).
23. M. Biyajima *et al.*, *Progr. Theor. Phys.* **84**, 931 (1990).
24. A. Białas, *Acta Phys. Pol.* B **23**, 561 (1992).
25. P. Bak, C. Tang and K. Wiesenfeld, *Phys. Rev. Lett.* **59**, 381 (1987); P. Bak and K. Chen, *Sci. Am.* **264**, 46 (1991).

26. G. Gustafson and A. Nilsson, *Z. Phys.* C **52**, 533 (1991).
27. W. Ochs and J. Wosiek, *Phys. Lett.* B **289**, 159 (1992); ibid. **305**, 144 (1993); *Z. Phys.* C **68**, 269 (1995).
28. Y.L. Dokshitzer and I.M. Dremin, *Nucl. Phys.* B **402**, 139 (1993).
29. Ph. Brax, J.-L. Meunier and R. Peschanski, *Z. Phys.* **C62**, 649 (1994).
30. B. I. Ermolaev and V. S. Fadin, *JETP Lett.* **33**, 269 (1981); A. H. Mueller, *Phys. Lett.* B **104**, 161 (1981); A. Bassetto, M. Ciafaloni, G. Marchesini and A. H. Mueller, *Nucl. Phys.* B **207**, 189 (1982); G. Marchesini and B. R. Webber, *Nucl. Phys.* B **238**, 1 (1984).
31. M. Acciari *et al.*, L3 Coll., *Phys. Lett.* B **353**, 145 (1995).
32. Ya. I. Azimov, Yu. L. Dokshitzer, V. A. Khoze and S. I. Troyan, *Z. Phys.* C **27**, 65 (1985).
33. B. Buschbeck, P. Lipa, F. Mandl in *Proc. 7th Int. Workshop on Multiparticle Production, "Correlations and Fluctuations", Nijmegen*, eds. R.C. Hwa *et al.* (World Scientific, Singapore, to be published).

Finite Size Effects In Hadronic And Nuclear Collisions[a]

Debapriyo Syam[a] and Sibaji Raha[b,c]

a) Presidency College, 86/1, College Street, Calcutta 700073, INDIA

b) Bose Institute, 93/1, Acharya Prafulla Chandra Road, Calcutta 700009, INDIA

c) Electronic Mail : sibaji@boseinst.ernet.in

ABSTRACT

We look at some hitherto unanalysed aspects of high energy hadronic collisions and discover novel finite size effects which induce fluctuations leading to approximate KNO scaling at low energy and also its violation at high energies. Extending these considerations to the case of heavy ions, we argue that a careful analysis of the data currently accessible should test the validity of our premise.

[a]This paper is a small token of respect and affection for Prof. Peter Carruthers on the occasion of his 61st. birthday.

High energy collision leading to multiparticle production has been at the forefront of research in particle physics for several decades now. Even though the consensus has emerged over the past decade that Quantum Chromodynamics (QCD) is the true theory of strong interactions, the non-perturbative aspects of low p_T processes, which constitute the bulk of multiparticle events, leave the theory almost unusable for these investigations. Phenomenological models, including some collective classical models, have therefore been customarily used, with varying degrees of success, to analyse such data. Unfortunately, the applicability of the collective models has thus far been limited only to *average* quantities (to wit, $< n >$,$< p_T >$,$< \frac{dn}{dy} >$ etc.) and the interesting physics that emerges from an event-by-event analysis remains unaddressed in this framework.

In the absence of a *tractable* microscopic theory, it is necessary to study the systematics of multiparticle production processes and look for scaling laws which may point the way to the underlying dynamics. The Feynman scaling [1], proposed in 1969, was short-lived. However, assuming the validity of Feynman scaling, Koba, Nielsen and Olesen [2] proposed in 1972 a new kind of scaling law, known in the community as KNO scaling, which was expected to be obeyed at asymptotic energies. Surprisingly, even though Feynman scaling was not observed, KNO scaling was found to describe the multiparticle data beginning already at ISR energies. It was at the SPS energies that deviations from KNO scaling were observed. Despite this deviation, it is obvious that KNO scaling is a proven feature of multiparticle production processes over a large energy range ($\sqrt{s} \sim 10 - 100$ GeV), the source of which is yet unknown.

There has been quite a few attempts at understanding the origin of KNO scaling, the notable among them being the negative binomial (NB) distribution approach of Carruthers and Shih [3], Giovanini and Van Hove [4], Biyajima [5] and others. In a nutshell, the NB distribution results from a convolution of multiplicity distributions from k independent chaotic sources, each with an average multiplicity $\frac{<n>}{k}$. The NB distribution provided a convenient *parametrization* of the multiplicity distribution data [6], with the empirical observation that k decreased with energy $\sqrt{s}$ - a somewhat counter-intuitive finding. Moreover, there exists, arguably, no *overwhelming* theoretical motivation, other than the limited phenomenological success, behind the adoption of the NB distribution. To this end, we introduce in this work an *heuristic* picture which yields rough KNO scaling in a *natural* way and more importantly, accommodates a violation of KNO scaling at higher energies in an equally natural fashion.

We start by adopting the viewpoint that a semi-classical, collective description [7] of multiparticle production in high energy hadronic collisions is, on the whole, valid. There however remains the problem [8] of the non-vanishing (partonic/hadronic) mean free path λ. In what follows, this will indeed prove to be the most important ingredient.

Perturbative QCD estimates [9,10] of qq, qg and gg cross sections indicate that the partonic mean free paths, while smaller than a typical hadronic size of $\sim 1fm$, are not very much smaller than $1fm$. In these circumstances, (*i.e.* $\lambda \not\ll d$, a typical hadronic length scale), we must expect, in addition to the geometrical (impact parameter) fluctuations and quantum statistical ('thermal') fluctuations, inherent fluctuations due to disorderly motion of the constituents which become manifest because of the finite size of the *fireball* and lead to a *fuzziness* of the surface. Confining our attention to the last named form of fluctuations, we suggest (see, *e.g.*, [11]) that the relative fluctuation $\frac{\delta f}{<f>}$ in any physical quantity f should be a function of ξ/d, where $\xi(\sim \lambda)$ is the scale of *surface irregularities* :

$$\frac{\delta f}{< f >} \sim \Phi_f\left(\frac{\xi}{d}\right) \tag{1}$$

The form of the function Φ for the physical variable f may be arbitrary, with the qualification that it vanishes in the limit $\xi/d \to 0$. Since d is largely independent of $\sqrt{s}$, the dependence of ξ/d comes in primarily through the variation of ξ with $\sqrt{s}$, which is expected to be small. We shall come back to this issue later on.

Without further ado, we can readily see how an *approximate* KNO scaling would emerge out of these considerations. The fluctuation δS in entropy would be,

$$\frac{\delta S}{< S >} \sim \Phi_S\left(\frac{\xi}{d}\right) \tag{2}$$

which, under the assumption of equilibrium, can be equivalently written as

$$\frac{\delta n}{< n >} \sim \Phi_S\left(\frac{\xi}{d}\right) \tag{3}$$

As long as $\Phi_S(\frac{\xi}{d})$ is independent (or at least a slowly varying function) of $\sqrt{s}$, KNO scaling follows. Deviation from KNO scaling requires a change in $\Phi_S(\frac{\xi}{d})$ with energy and the broadening of the KNO curve in the SPS data [6] implies that $\Phi_S(\frac{\xi}{d})$ increases with $\sqrt{s}$.

We are not aware of any prescription for evaluating the function $\Phi_S(\frac{\xi}{d})$. Till now, our considerations have been most general, that is, irrespective of any physical model of the collision process. But it is obvious that the form of the function Φ_S must be governed by the hadronisation mechanism (which, as we discussed above, is as yet uncalculable in a microscopic theory). For a coarse-grained understanding of the form of $\Phi_S(\frac{\xi}{d})$, we therefore take recourse to the following *heuristic* model, where the total surface area of the *fireball* of volume V (as revealed by BE correlation measurements) is denoted by Σ. BE correlation measurements presume a specific shape for fixing the parameters, but the shape of the *fireball* does not play a very important role in what follows.

The fluctuations due to non-zero ξ/d can then be visualised as giving rise to *undulations* in the *fuzzy* surface of the hadronisation volume V- see figure 1 for a schematic picture. Clearly excessive amounts of undulations is unlikely, a feature we have consciously incorporated in our model.

Now, in the spirit of collective models, we can estimate the variation in entropy due to these surface fluctuations. The average total entropy $< S >$ in the *fireball* is $\sim T^3V$, whereas the variation of temperature with volume (for a given $\sqrt{s}$) may be parametrized as

$$T \sim V^{-\beta}, 0 < \beta < \frac{1}{4} \tag{4}$$

the two limits corresponding to the Pomeranchuk-Landau and the static (*i.e.*, no collective flow) fireball ansätze, respectively, as two extreme cases. Thus we can write $< S > \sim V^{1-3\beta}$, so that

$$\frac{\delta n}{< n >} \approx \frac{\delta S}{< S >} \sim (1 - 3\beta)\frac{\delta V}{V} \tag{5}$$

We, therefore, need to estimate the volume fluctuation δV corresponding to fluctuations of size ξ in the surface. With the passage of time, the volume of the *fireball* may be larger or smaller than the value arrived at from the hydrodynamic picture, but the r.m.s. value of volume fluctuation will be non-zero. The r.m.s. value in question may be estimated within a simple model. For concreteness, we assume in the following exercise that the *average* volume is cylindrical in shape.

If, in an event there are ν_+ outward *bumps* and ν_- inward *dents* on the surface (*cf.* fig. 1), then the probability of such a fluctuation may be written as $P \sim exp(\chi)$, where

$$\chi \approx -\frac{\alpha \cdot (\nu_+ + \nu_-) \cdot \xi^2}{\Sigma} - \frac{(\nu_+ - \nu_-)^2}{4\bar{\nu}} \tag{6}$$

where $\bar{\nu}$ is the average number of *bumps* or *dents* in an event. The first term in (6) corresponds to the relative change in the surface area of the *fireball* due to the fluctuation. The constant α is small, as the cost in terms of energy of producing one extra bump or dent in a fuzzy and undulated surface should be small. The second term is the usual exponent found in the 'coin-tossing' problem of statistics. The first term immediately implies $\bar{\nu} \propto \Sigma$. The second term gives $< (\nu_+ - \nu_-)^2 >= 2\bar{\nu} \propto \Sigma$. Now each bump or dent originating from a fluctuation of scale ξ possesses an area of $\sim \xi^2$ so that $\bar{\nu} \sim \Sigma/\xi^2$. Thus the standard deviation $\delta\Sigma$ in the surface area is given by $\delta\Sigma \sim< (\nu_+ + \nu_-) > \xi^2 \sim \Sigma$, while the corresponding fluctuation in volume δV is :

$$\delta V \sim< (\nu_+ - \nu_-)^2 >^{1/2} \xi^3 \sim (\Sigma/\xi^2)^{1/2}\xi^3 \sim \xi^2\Sigma^{1/2} \tag{7}$$

Thus, $\frac{\delta V}{V} \sim \frac{\xi^2\sqrt{\Sigma}}{V}$. Now $V(\approx d\Sigma)$, as extracted from BE correlation measurements over a large energy range ($\sqrt{s} \sim 30$ - 900 GeV), appears to be rather insensitive [12] to $\sqrt{s}$. We thus have

$$\frac{\delta n}{< n >} \sim \frac{1 - 3\beta}{d\sqrt{\Sigma}}\xi^2 \propto \xi^2 (approx) \tag{8}$$

which is expected to be a slow function of $\sqrt{s}$.

Broadening of the KNO curves with $\sqrt{s}$ requires that ξ increase with $\sqrt{s}$. That such a change is indeed possible and that it is even in quantitative agreement with the observed broadening emerge from the following simple physical picture employing straight-forward dynamical arguments.

In a thermal system, the dependence of λ on T may be parametrized as $\lambda \sim T^\delta$. If the system comprises only asymptotically free point-like partons, then $\delta \approx -1; [\lambda \sim (n\sigma)^{-1} \sim T^{-1}; n \sim T^3; \sigma \sim T^{-2}]$. The limits on δ should therefore be $-1 \le \delta \le 0$. Now, if the average energy per particle in the thermal system is $< \epsilon >$ $(\sim T)$, we can write, for the scale of fluctuation ξ,

$$\xi \approx \left[\frac{< \epsilon >}{t\lambda}\right]^\mu \cdot \lambda \sim< \epsilon >^\mu \cdot \lambda^{1-\mu}$$

where μ is a parameter and t the string tension. Usually, a fluctuation is generated by a parton moving out of the core region. For a random-walk-type excursion of this parton, $\mu \approx \frac{1}{2}$ and for collisionless motion, $\mu \approx 1$. One should therefore expect that $\frac{1}{2} \le \mu \le 1$, with the appropriate value being close to $\frac{1}{2}$ for a thermal system of asymptotically free particles.

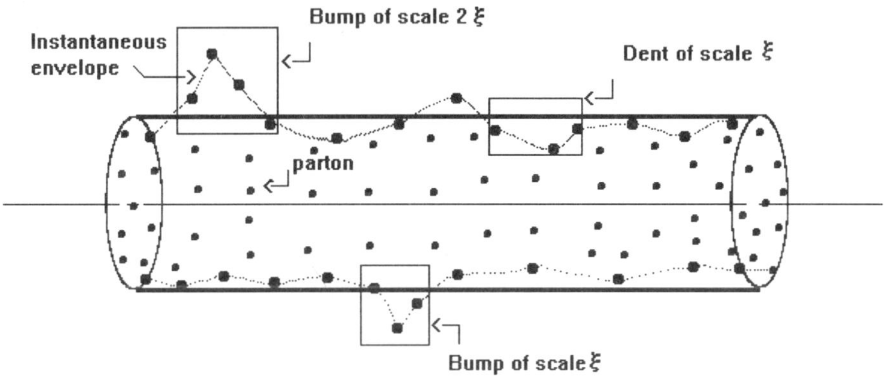

Figure 1: Schematic Picture of the hadronisation volume

The hadronisation energy $W = k\sqrt{s}$ is related to T as $W \sim VT^4 \sim T^4$ (assuming a fixed flow pattern); thus for constant k (we have ignored the possibility of changing inelasticity at the present stage), $T \sim (\sqrt{s})^{1/4}$. Using this in the expression for ξ, we have,

$$\xi \sim (\sqrt{s})^{[\mu+(1-\mu)\delta]/4} \tag{9}$$

so that from (8)

$$\frac{\delta n}{<n>} \propto \xi^2 \sim (\sqrt{s})^{[\mu+(1-\mu)\delta]/2} \equiv (\sqrt{s})^\gamma \tag{10}$$

Clearly $\gamma = 0$ ($\mu = \frac{1}{2}, \delta = -1$) would give exact KNO scaling. For the range of values of μ and δ, $0 \leq \gamma \leq \frac{1}{2}$ although the more reasonable value to expect ($\delta \approx -1, \mu \approx \frac{1}{2} + \epsilon$) is closer to 0, since the wisdom of QCD leads us to expect that the system should behave as largely free for a major part of the early evolution.

As there exists no well-defined, dynamical prescription for computing γ, we set out to determine it from experimental data [13,14]. (Ideally, this ought to be done only with data for energies below the SPS energy, *i.e.*, the ISR data alone. However, for reasons that will become *a posteriori* clear (cf. fig. 2), we have included also the data at $\sqrt{s} = 200$ GeV for the determination of γ.) We extract $\frac{D_2}{<n>}|_{F.S.}$ (F.S. stands for finite size) from the experimental D_2 and $<n>$ ($\equiv <n>_{ch}^{ND} -2$, ch refers to charge, ND to non-single-diffractive and the 2 accounts for the leading particles) by assuming that D_2 for thermal fluctuations behaves like $<n>^{1/2}$ (and ignoring the impact parameter fluctuation for the time being) so that $\frac{D_2}{<n>}|_{F.S.} = \sqrt{\left(\frac{D_2}{<n>}|_{exp}\right)^2 - \frac{1}{<n>}}$. An *eyeball* fit yields $\frac{D_2}{<n>}|_{F.S.} \approx (\sqrt{s})^{0.1}$ (cf. below). Taking $\gamma = 0.1$, we can then compute $\frac{D_2}{<n>}|_{F.S.}$ at the higher SPS energies ($\sqrt{s} = 546$ GeV and 900 GeV) in a straightforward manner and compare them with the corresponding experimental quantities, as shown in fig. 2. The agreement is obviously very good. A *least squares* fit for all the data points shown in fig. 2 also yields $\gamma = 0.0997$ (with χ^2/NDF = 0.325), a number sufficiently small to vindicate the above arguments, and sufficiently close to the value extracted from the eyeball fit. Extrapolating to the Tevatron energy of $\sqrt{s} = 1.8$ TeV, one obtains $\frac{D_2}{<n>}|_{F.S.} = 0.6320$ which translates to, for the preliminary value of $<n>_{ch}^{ND} \approx 44$, $\frac{D_2}{<n>}|_{exp} \approx 0.6506$. This may be considered an approximation to $\frac{D_2}{<n>}|_{exp}$ as we have neglected impact parameter fluctuation as well as inelasticity fluctuation. Examination of fig. 2 also reveals that for $\sqrt{s} \geq 120$ - 150 GeV, violation of KNO scaling (the dashed line in fig. 2) should be visible, if the data sample is large enough, *i.e.*, the error bars are small enough.

At this point, we feel it important to examine the validity of our premise that in high energy hadronic collisions, a collective description is, on the whole, tenable (see above). In particular, the increased occurrence of *minijets* with increasing $\sqrt{s}$ may be posed as a counter-example. We shall however argue here that the present model of fluctuations, curiously enough, can accommodate these *minijets* in a natural fashion, as the scale of fluctuations ξ increases with $\sqrt{s}$. This comes about in the following manner.

A high energy parton is likely to produce a *spike* as it leaves the core of the *fireball*. As the *spike* grows, it may fragment into several pieces (mesons). The probability P_{sc} of such a scission of the spike should naturally be larger for greater energy ϵ of the parton associated with the spike and lower for greater string tension t. Moreover, the narrower the spike, the greater the probability of scission, so that P_{sc} should decrease with increasing $\mathcal{A}$, the cross sectional area of the spike. Drawing inspiration from the expression for the flux tube fission probability in Schwinger's 1-D field theoretic model, one can parametrize P_{sc} as,

$$P_{sc} \approx 1 - exp\left[-K\left(\frac{\epsilon}{t\sqrt{\mathcal{A}}}\right)^{\Delta}\right], \Delta > 0$$

where K and Δ are parameters. Recalling that $\epsilon \sim T \sim W^{1/4}$ and $\sqrt{\mathcal{A}} \sim \xi \sim W^{1/20}$, we can write $P_{sc} \approx 1 - exp[K'(W)^{\Delta'/5}]$. Clearly, P_{sc} is proportional to the number of events with minijets at any particular $\sqrt{s}$, within the framework of our scanario.

K' and Δ' may be estimated from the available data, that, from ISR to the Collider, the number of events with minijets increases from 2 - 5% to 15 - 17%. One can then easily calculate P_{sc} at the Tevatron energy which turns out to be 30 - 45%. Comparing this with the experimental observation that at $\sqrt{s} = 1.8$ TeV, almost half of the events are *semi-hard*, it is very tempting to claim that minijets are a natural consequence of the finite size fluctuations discussed here.

Accepting the possibility of fluctuations larger than the fundamental scale ξ allows us to extend these considerations to make other predictions which are falsifiable with the present generation of accelerators. For the case of equal heavy ions of mass number A, we may rewrite (in a discretized approximation) the probability of fluctuations (eq. 6) as :

$$\begin{aligned} P &\sim exp(\chi), \ where, \\ \chi &\approx -\sum_m \left[\frac{\alpha \cdot (\nu_+ + \nu_-)_m \cdot \xi_m^2}{\Sigma} + \frac{(\nu_+ - \nu_-)_m^2}{4\bar{\nu}_m}\right] \end{aligned} \tag{11}$$

where $\xi_m \equiv m\xi, m = 1, 2, \ldots,$. In the above, $(\nu_\pm)_m$ is the number of bumps (dents) of scale ξ_m. Let us recall that the *average* surface area Σ scales as $A^{2/3}$ and the volume V as A. For large Σ, many values of m will become operative. Then, considering the first term in (9), for any small m, we find that the value of $(\nu_+ + \nu_-)_m$ for which $\alpha \cdot (\nu_+ + \nu_-)_m \cdot \xi_m^2/\Sigma$ is comparable to unity scales as ξ_m^{-2}. Hence, one can conclude that

$$\frac{< \nu_+ + \nu_- >_m}{< \nu_+ + \nu_- >_{m=1}} \approx m^{-2} \tag{12}$$

so that

$$\sqrt{< (\nu_+ - \nu_-)_m^2 > / < (\nu_+ - \nu_-)_{m=1}^2 >} \approx m^{-1} \tag{13}$$

implying that as long as $\Sigma \gg \alpha\xi_m^2$, the *unevenness* of the surface due to fluctuations of scale ξ_m are as important as those for scale ξ. It is also clear that the mean scale size $< \xi_m >$ grows like $\Sigma^{1/4}$ and $< (\nu_+ + \nu_-)_{<\xi_m>} >\sim \sqrt{\Sigma}$.

Thus, the volume fluctuation δV is estimated to be

$$\delta V \sim \sqrt{< (\nu_+ - \nu_-)_{<\xi_m>}^2} < \xi_m >^3 \sim \Sigma^{1/4}\Sigma^{3/4} \sim \Sigma \sim A^{2/3} \tag{14}$$

so that $\delta V/V \sim A^{-1/3}$.

(Generalising further, one may consider a continuous distribution of fluctuation sizes ξ' and write

$$\chi \approx -\int_0^\infty \left[\frac{\alpha\{\nu_+(\xi')+\nu_-(\xi')\}\xi'^2}{\Sigma} + \frac{\{\nu_+(\xi')-\nu_-(\xi')\}^2}{4\bar{\nu}(\xi')}\right] d\xi' \tag{15}$$

whence similar conclusions can be drawn.)

We therefore conclude that the finite size induced fluctuations as discussed here lead, in AA collisions, to $\delta n \propto < n >$ (*viz.* eq. 5), with the constant of proportionality decreasing as $A^{-1/3}$. This then immediately suggests that if, for high multiplicity and large E_T selected events (so as to minimise the effect of impact parameter fluctuations), we plot KNO curves for the multiplicity distribution for various A-s at the same cm energy per nucleon, the KNO curves should show a narrowing trend, albeit small ($\sim A^{-1/3}$), for increasing mass number A. Such an analysis is well within the reach of presently available experimental set-ups and would go a long way toward verifying (or, equally importantly, falsifying) the *heuristic* picture described here.

Multiplicity fluctuation measurements at CERN energies for *light* heavy ions (O^{16} and S^{32}) have been reported by various collaborations. Of particular interest in the present context are the NA35 data [15,16] for $S + S$ at 200 GeV per nucleon, corresponding to $\sqrt{s} \sim$ 20 GeV. These authors report that for *central* collisions, $C_2 - \frac{1}{<n>} = (\frac{D_2}{<n>}|_{exp})^2 + 1 - \frac{1}{<n>}$ is 1.03 ± 0.01, which yields $\frac{D_2}{<n>}|_{F.S.} = \sqrt{0.03 \pm 0.01} \approx 0.17 \pm 0.03$. This value is to be compared with the $\frac{D_2}{<n>}|_{F.S.}$ estimated from the corresponding hadronic case $\approx 0.44 \cdot A^{-1/3} \approx 0.14$. Given the crudeness of the $A^{-1/3}$ approximation for this case and the ambiguities over the criteria for centrality, the agreement, to say the least, is remarkable.

As a further preliminary test of our scenario, let us look at the data for $^{16}O + Ag$ and $^{32}S + Ag$ of the HELIOS/2 [17] Collaboration. For central events, they quote $\frac{D_2}{<n>}|_{F.S.} \approx 0.25 \pm 0.04$ for ^{16}O and 0.14 ± 0.04 for ^{32}S, a significant decrease with A. Noting that the second order factorial moment F_2 is equal to $\left(\frac{D_2}{<n>}|_{F.S.}\right)^2 + 1$, we can also study the effect A_{target} with the help of the EMU01 data [18]. They found $F_2 = 1.09 \pm 0.03$ for $^{32}S + Ag$ and 1.04 ± 0.01 for $^{32}S + Au$, which translate to $\frac{D_2}{<n>}|_{F.S.} \approx 0.3 \pm 0.05$ and 0.2 ± 0.03, for the Ag and Au targets, respectively. In both these cases, the decrease in $\frac{D_2}{<n>}|_{F.S.}$ with increasing mass number is clearly found in the data, as predicted by our model. Obviously, a quantitative comparison with the predicted $A^{-1/3}$ behaviour is premature, given the ambiguities over the centrality criteria, estimation of total number of nucleons involved etc., but we hope these observations will motivate detailed study of various central AA processes at the same energy.

To summarize, we argue that fluctuations due to finite size effects provide a natural basis for the origin of the observed KNO scaling and violations thereof in hadronic collisions. This would solve the long-standing puzzle of why KNO scaling, predicted to be an asymptotic phenomenon by the original authors, appears to occur at low energies but is violated at higher energies. In the case of relativistic heavy ions, we show that these fluctuations may have measurable consequences.

References

1. R. P. Feynman, *Phys. Rev. Lett.* **23**, 1415 (1969)
2. Z. Koba, H. B. Nielsen and P. Olesen, *Nucl. Phys.* **B40**, 317 (1972)
3. P. Carruthers and C. C. Shih, *Phys. Lett.* **127B** , 242 (1983) ; **137B**, 425 (1984)
4. A. Giovanini and L. Van Hove, *Z. Phys.* **C30** , 391 (1986)
5. M. Biyajima, *Prog. Theor. Phys.* **69**, 966 (1983)
6. R. E. Ansorge *et al.* (UA5 Collaboration), *Z. Phys.* **C43**, 357 (1985); G. J. Alner *et al.* (UA5 Collaboration), *Phys. Lett.* **167B**, 476 (1986)
7. P. Carruthers, *Ann. N. Y. Acad. Sci.* **229**, 91 (1974)
8. E. L. Feinberg, in : *Local Equilibrium in Strong Interaction Physics* (eds. D. K. Scott and R. M. Weiner), p. 15, World Scientific, Singapore (1985) ; L. McLerran, *ibid.*, p. 70
9. E. V. Shuryak, *Phys. Rep.* **61**, 71 (1980)
10. P. Danielewicz and M. Gyulassy, *Phys. Rev.* **D31**, 53 (1985)
11. K. Huang, *Statistical Mechanics*, p. 104, John Wiley, New York (1987)
12. T. Kageya *et al.*, in : *Correlations and Multiparticle Production* (eds. M. Plümer, S. Raha and R. M. Weiner), p. 92, World Scientific, Singapore (1991)
13. A. Breakstone *et al.*, *Phys. Rev.* **D30**, 528 (1984)
14. C Albajar *et al.* (UA1 Collaboration), *Nucl. Phys.* **B309**, 405 (1988)
15. J. Bartke, E. Gladysz-Dziadus and P. Stefanski, in : *Correlations and Multiparticle Production* (eds. M. Plümer, S. Raha and R. M. Weiner), p. 198, World Scientific, Singapore (1991)
16. J. Bartke, E. Gladysz-Dziadus and P. Stefanski, *Nucl. Phys.* **A545**, 321c (1992)
17. T. Akesson, *Proc. 24th Int. Conf. on High Energy Physics*, Munich, p. 1382 (1988)
18. M. I. Adamovitch *et al.* (EMU01 Collaboration), *Phys. Lett.* **242B**, 512 (1990)

SIMPLE MODEL FOR TOTAL CROSS SECTIONS[1]

Jean-René Cudell
Inst. de Physique, U. de Liège, Bât. B-5, Sart Tilman, B4000 Liège, Belgium
cudell@gw.unipc.ulg.ac.be
Kyungsik Kang[2]
Physics Department, Brown University, Providence, RI 02912, U.S.A.
kang@het.brown.edu
Sung Ku Kim[3]
Dept. of Physics, Ewha Women's University, Seoul 120-750, Korea
skkim@mm.ewha.ac.kr

(presented by Kyungsik Kang)

Abstract

Adopting the philosophy *à la* Donnachie and Landshoff that simple pole exchanges could account for all data of total, elastic and diffractive scattering cross sections to present energies, we show that such simple pole fits to pp and $\bar{p}p$ total cross sections are indeed very successful. We assess the uncertainties of the various parameters by making careful statistical analysis of the data and their correlations. In particular, the pomeron intercept which controls total cross sections and the real part of the elastic amplitude at high energies is shown to lie anywhere between 1.07 and 1.11, with a preferred value 1.096.

[1]presented at the Workshop on "The State of Physics at the End of the 20th Century", in honor of Pete Carruthers' 61st Birthday, Santa Fe, NM, 26-29 October 1996.

[2]supported in part by the US DOE contract DE-FG02-91ER 40688-Task A.

[3]supported in part by the Korea Science and Engineering Foundation through the Center for Theoretical Physics, Seoul National University, by the Ministry of Education through contract BSRI-95-2427, and by the Non-Directed Research Fund of Korea Research Foundation.

It is well-known that total cross sections rise at high energy. The simplest object responsible for the rising behavior would be a Regge trajectory, the pomeron, with an intercept somewhat greater than 1. This object has been hypothesized a long time ago [1], and presumably arises from the gauge sector of QCD. Its intercept and slope are then fundamental numbers characterizing the pure gauge sector of Yang-Mills SU(3).

There has been a renewed interest in the pomeron after the observation of rapidity gaps in deep inelastic *ep* scattering at HERA [2] and the suggestion that such cross sections might be used for the detection of new physics [3]. One can view the emergence of gaps as resulting from the emission of a pomeron by the colliding proton (which then remains in a color-singlet state) followed by a pomeron-photon collision. The validity of this picture, as well as the measurement of the pomeron intercept and structure function is thus a central issue for HERA, and will no doubt have a bearing on the extrapolation to the physics at the LHC energies. Hence we propose to re-evaluate the pomeron intercept from the simple-pole model fits to the total cross sections as carefully as possible and to estimate the errors on the various parameters. The information on the soft pomeron intercept comes from the high energy behavior of the total cross sections. Because of the presence of the sub-leading Regge poles at low energy and because the unitarity effect due to multipomeron exchanges at high energy, one must not only determine the best parameterization but also the range of validity of the model. It is clear that a simple minded χ^2 test cannot be sufficient, primarily because the cross section data contain many points that are inconsistent with their neighboring points. We therefore must invent a reasonable method to filter the data sets independently of any underlying theoretical model or prejudice. A reasonable criterion is that a given data point should not deviate by more than 1 or 2σ from the average of all data in a bin of ± 1 GeV centered around it and yet the central values of the parameters and their errors should not depend too sensitively on the filtering itself. The stability of the parameter values is more reasonable criterion than the value of ${\chi^2}_{min}$ in our opinion. In the following, we use two strategies to evaluate the best central values and their errors. The first is to use all the data available [4, 8]. This gives us the central values. However, some of the data points are incompatible at the 2σ level or more so that the value of χ^2 will be artificially inflated. In fact the best fits [9, 10] that one can produce have a $\chi^2/d.o.f.$ of 1.3 or more. We then use the data sets that are filtered by using the proposed criterion, which will give us stable central values of the parameters and their uncertainties. The best selection criterion would be based on physics arguments to separate wrong experimental results. Unfortunately, this would prove infeasible for old (ISR) data, where most of the incompatibilities lie. We give in Table 1 the number of data points that are kept before and after filtering. The full data sets are available at http://nuclth02.phys.ulg.ac.be/Data.html.

data set	σ_{tot}^{pp} (mb)	$\sigma_{tot}^{\bar{p}p}$ (mb)	ρ^{pp}	$\rho^{\bar{p}p}$
P.D.G. [4]	94	28	-	-
P.D.G. [4] - 2σ	84	28	-	-
P.D.G. [4] - 1σ	65	20	-	-
Ref. [8]	66	29	41	13
Ref. [8] - 2σ	60	28	38	13
Ref. [8]- 1σ	53	19	31	13

Table 1: The number of points kept after data selection, for $\sqrt{s} > 10$ GeV.

Note that the 2σ selection includes both CDF and E710 points, whereas the 1σ one rejects them both. This procedure is not unlike the one followed by UA4/2 in [7]. As the value of χ^2 - χ^2_{min} is distributed as a χ^2 with N parameters of the model, the $\Delta\chi^2$ corresponding to 70% confidence level(C.L.) is 6.06 [11, 12] in the DL case with 5 parameters. The errors we quote in this paper are to this χ^2-interval, to be contrasted to those quoted in the Particle Data Group(PDG) [10] who simply renormalized the χ^2 to $\chi^2/d.o.f. = 1$ and let the new χ^2 change from the minimum by one unit.

Donnachie and Landshoff (DL) have proposed[5] to fit the pp and $\bar{p}p$ cross section using a minimal number of trajectories: the leading meson trajectories of the degenerate a/f (C=+1) and ρ/ω (C=-1), plus the pomeron trajectory. They fit data for $s > 100$ GeV^2, as lower trajectories would then contribute less than 1%, which is less than the errors on the data. The result of their fit [6] is a pomeron intercept of 1.0808, for which they did not quote a χ^2 or error bars and said that the χ^2 was very flat near the minimum. We show in Table 2 our results for such a fit. We use the usual definition

$$\chi^2 = \sum_i \left(\frac{d_i - s_i^{-1} \mathcal{I}m\mathcal{A}(s_i, 0))}{e_i} \right)^2 \tag{1}$$

parameter	all data	filtered data (2σ)	filtered data (1σ)
χ^2	410.8	80.3	32.4
χ^2 per d.o.f.	3.16	0.62	0.25
pomeron intercept-1	$0.0912^{+0.0077}_{-0.0070}$	$0.0887^{+0.0079}_{-0.0071}$	$0.0863^{+0.0096}_{-0.0084}$
pomeron coupling (mb)	$19.3^{+1.5}_{-1.7}$	$19.8^{+1.6}_{-1.7}$	$20.4^{+1.8}_{-2.1}$
$\rho/\omega/a/f$ intercept-1	$-0.382^{+0.065}_{-0.071}$	$-0.373^{+0.067}_{-0.073}$	$-0.398^{+0.083}_{-0.090}$
ρ/ω coupling $C_-(pp)$ (mb)	$-13.2^{+4.1}_{-6.5}$	$-13.3^{+4.3}_{-6.9}$	$-15.3^{+5.7}_{-10.1}$
a/f coupling (mb)	69^{+20}_{-13}	62^{+19}_{-12}	67^{+25}_{-16}

Table 2: Simple pole fits to total pp and $\bar{p}p$ cross sections, assuming degenerate $C = +1$ and $C = -1$ exchanges.

with $d_i \pm e_i$ for the measured pp or $\bar{p}p$ total cross section at energy $\sqrt{s_i}$, and

$$\mathcal{I}m\mathcal{A}(s,0) = C_- s^{\alpha_m} + C_+ s^{\alpha_m} + C_P s^{\alpha_P} \tag{2}$$

where C_- flips sign when going from pp to $\bar{p}p$. We use the same data set as DL [4] to determine the central value, and use the selected data sets at the 1- or 2-σ level to determine the errors. We see that the fit to all data gives a totally unacceptable value, $\chi^2 = 410$ for 135 data points with 5 parameters, corresponding to a C.L. of 2×10^{-36} ! There are two possible outcomes to such a high value of the χ^2: either the model is to be rejected, or some of the data are wrong. As we already mentioned, there are a few obviously wrong points within the data. Hence before rejecting the model, let us eliminate those points. Table 2 shows that the central values and their errors indeed do not depend too much on the filtering itself, while filtering the data does change the value of χ^2_{min} drastically so that the model becomes perfectly acceptable. Also we see that the pomeron intercept is determined to be about 1.090, and that it could be as high as 1.096. We think the stability of the parameter values is a more important than the value of χ^2_{min} itself.

As for the energy range of validity of the model, we require the two basic requirements: that the $\chi^2/d.o.f.$ be of the order of 1, and that the determination of the intercept be stable. We show in Fig. 1 the result of varying the energy range. Clearly, the lower trajectories seem to matter for $\sqrt{s}_{min} < 10$ GeV, whereas the upper energy does not seem to modify the results (in other words, there is no sign of the onset of unitarisation). Hence we adopt $\sqrt{s}_{min} = 10$ GeV as the lowest energy at which the model is correct.

This happens to be the point at which χ^2_{min} is lowest. This dependence on the lower energy cut explains why both the PDG [10] and Bueno and Velasco [13] obtain an wrong value for the intercept, much lower than ours.

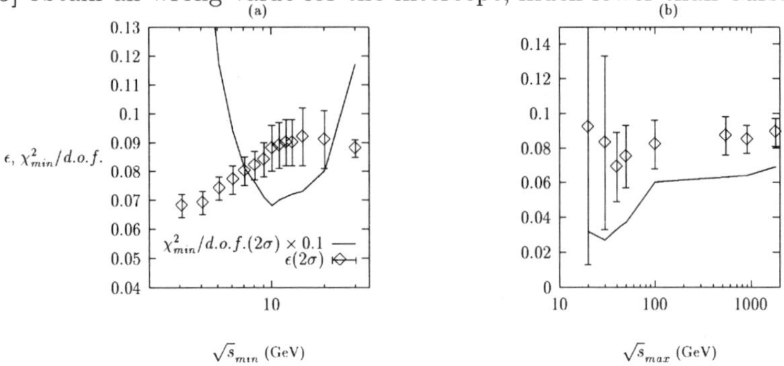

Figure 1: DL intercept-1 as a function of the lower (a) and upper (b) energy cuts on the data. The curve shows the $\chi^2/dof.$ for data filtered at the 2σ level.

One must wonder if it is possible to get a better determination of the soft pomeron intercept by using more data. The PDG [10] obtained very narrow determinations of the pomeron intercept from the other hadronic reactions. We believe that their conclusions are wrong, and illustrate this in the case of the $\pi^{\pm}p$ total cross sections, for which they use $\sqrt{s}_{min} \approx 4$ GeV and obtain an intercept of 1.079 ± 0.003. We show in Fig. 2 our results for such a fit: for the data set filtered at the 2σ level (92 points), and for $\sqrt{s}_{min} = 4$ GeV, we obtain $\alpha_0 = 1.115^{+0.030}_{-0.023}$. Fig. 2(b) shows our best fit together with that of the PDG [10]. Although according to their estimate our central value for the intercept is 10 standard deviations from theirs, we see that the two fits are indistinguishable. Hence we believe that both their standard deviations and their central values are wrong. These conclusions are not affected by the use of the full data set instead of the one filtered at the 2σ level. The above value of the intercept in fact gives a slightly smaller $\chi^2/d.o.f.$ than the one quoted in the PDG [10]. Note that this intercept is consistent with the one we got from the analysis of pp and $\bar{p}p$ total cross sections. The conclusion from this exercise is that the errors from the low-energy hadronic data are large, especially if we use a low-energy cutoff of the order of 10 GeV. Hence we want to limit ourselves to pp and $\bar{p}p$ amplitudes at $t = 0$.

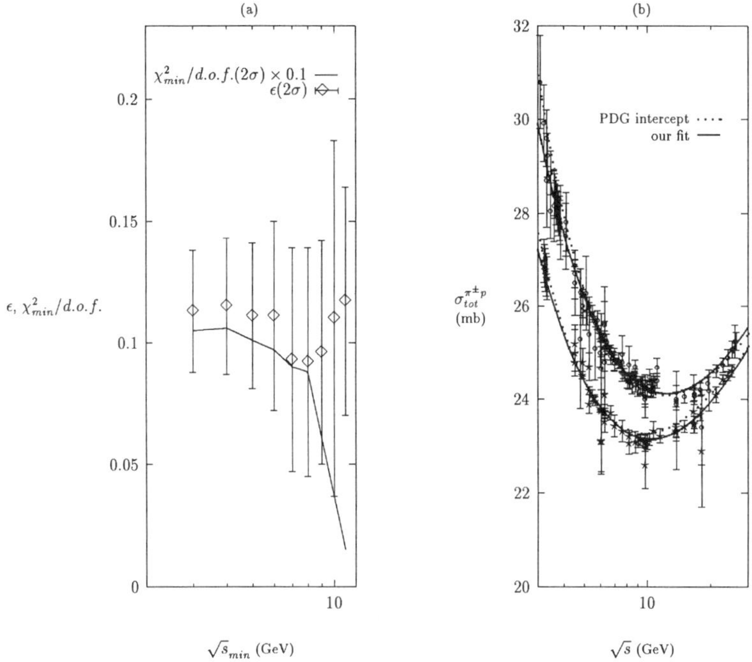

Figure 2: (a) shows the pomeron intercept from πp data as the lower energy cut on the data is changed; (b) shows our best fit to the data set filtered at the 2σ level together with that of the Particle Data Group.

One more piece of pp and $\bar{p}p$ data can be used however: the knowledge of the intercept is sufficient to determine the value of the real part of the amplitude, using crossing symmetry, and hence the measurements of the ρ parameter provide an extra constraint. We use the data collected in Ref. [8], and obtain a somewhat worse fit, as shown in Table 3, even when filtering data at the 1 or 2σ level. For the 2σ filtering of the data, the C.L. goes from 99.4% to 36%. We show the curves corresponding to the second column of Table 3 in Fig. 3 with dotted lines. Whether one should worry about the change of χ^2 and the change of central value for the parameters, is a matter of taste. But it is this small change of central values, combined with the effect of too low an energy cut, that lead Bueno and Velasco [13] to conclude that simple-pole parametrisations were disfavored.

parameter	all data	filtered data (2σ)	filtered data (1σ)
χ^2	561.3	168.3	94.9
χ^2 per d.o.f.	3.28	1.07	0.77
pomeron intercept−1	0.0840 ± 0.0050	$0.0817^{+0.0055}_{-0.0053}$	$0.0804^{+0.0064}_{-0.0061}$
pomeron coupling (mb)	20.8 ± 1.1	21.4 ± 1.1	21.8 ± 1.3
$\rho/\omega/a/f$ intercept−1	$-0.408^{+0.032}_{-0.033}$	$-0.421^{+0.034}_{-0.036}$	$-0.431^{+0.037}_{-0.040}$
ρ/ω coupling $C_-(pp)$ (mb)	$-14.0^{+2.6}_{-3.3}$	$-16.5^{+3.3}_{-4.2}$	$-17.7^{+3.7}_{-4.9}$
a/f coupling (mb)	$67.0^{+7.6}_{-6.7}$	$66.6^{+8.3}_{-7.2}$	$67.6^{+9.0}_{-7.8}$

Table 3: Simple pole fit to total pp and $\bar{p}p$ cross sections, and to the ρ parameter, assuming degenerate $C = +1$ and $C = -1$ meson exchanges.

Before concluding on the best value of the intercept, we need to examine the influence of low energy cut on the determination of the intercept. Although the energy cut $\sqrt{s}_{min}$ eliminates sub-leading meson trajectories, there is still an ambiguity in the treatment of the leading meson trajectories. In fact, a slightly different treatment to that of DL leads to a better χ^2 and to more stable parameters. Indeed, there is neither theoretical nor experimental reason to assume that the ρ, ω, f and a trajectories are degenerate. On the other hand, the data are not constraining enough to determine the effective intercepts of the four meson trajectories together with the pomeron intercept. We adopt an intermediate approach, which is to assume the exchange of separate + and − trajectories with independent intercepts:

$$Im\mathcal{A}(s,0) = C_- s^{\alpha_-} + C_+ s^{\alpha_+} + C_P s^{\alpha_P} \qquad (3)$$

The resulting numbers are shown in Table 4, and are plotted in Fig. 3 with plain lines. The χ^2 is smaller and the parameters are more stable than in the previous case. The bounds on the soft pomeron intercept hardly depend on the criterion used to filter the data, and intercepts as large as 1.108 are allowed. In order to better understand the treatment of the errors, we give in Table 5 the result of a fit to the data of Ref. [8]. We see that the results are very stable, especially those for the pomeron intercept, independently of the data filtering.

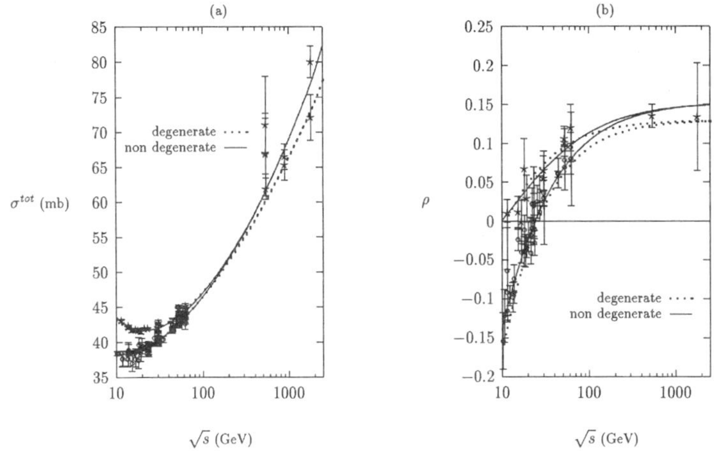

Figure 3: Best fits to 2σ filtered data. The dotted lines correspond to the original DL model given in Eq. (2), whereas the plain ones correspond to a model where the degeneracy of the lower trajectories is lifted, as in Eq. (3). The data points are the PDG data filtered at the 2σ level.

parameter	all data	filtered data (2σ)	filtered data (1σ)
χ^2	505.4	119.6	57.6
χ^2 per d.o.f.	2.99	0.77	0.47
pomeron intercept−1	$0.0990^{+0.0099}_{-0.0088}$	$0.0964^{+0.0115}_{-0.0091}$	$0.095^{-0.013}_{+0.010}$
pomeron coupling (mb)	$17.5^{+1.9}_{-2.0}$	$18.0^{+2.0}_{-2.2}$	$18.2^{+2.3}_{-2.6}$
ρ/ω intercept−1	$-0.494^{+0.056}_{-0.066}$	$-0.498^{+0.057}_{-0.067}$	$-0.510^{+0.064}_{-0.077}$
ρ/ω coupling $C_-(pp)$ (mb)	$-24.0^{+6.8}_{-10.9}$	$-26.5^{+7.7}_{-12.5}$	$-28.2^{+8.9}_{-15.4}$
a/f intercept−1	$-0.312^{+0.051}_{-0.052}$	-0.315 ± 0.058	$-0.324 + 0.066$
a/f coupling (mb)	$56.8^{+8.1}_{-6.7}$	$54.9^{+9.0}_{-7.2}$	$56.2^{+9.9}_{-7.8}$
$\sigma_{tot}(1.8$ TeV) (mb)	$77.6^{+2.5}_{-2.7}$	$76.8^{+2.9}_{-2.7}$	$76.4^{+3.4}_{-3.1}$
$\sigma_{tot}(10$ TeV) (mb)	$108.4^{+7.0}_{-6.7}$	$106.4^{+7.9}_{-6.7}$	$105.4^{+9.2}_{-7.5}$
$\sigma_{tot}(14$ TeV) (mb)	$115.8^{+8.3}_{-7.7}$	$113.5^{+9.3}_{-7.8}$	$112.3^{+10.8}_{-8.6}$

Table 4: Simple pole fit to total pp and $\bar{p}p$ cross sections, and to the ρ parameter, with non-degenerate $C = +1$ and $C = -1$ meson exchanges.

parameter	all data	filtered data (2σ)	filtered data (1σ)
χ^2	197.9	107.7	56.3
χ^2 per d.o.f.	1.39	0.82	0.52
pomeron intercept−1	$0.0955^{+0.0097}_{-0.0083}$	$0.0940^{+0.0092}_{-0.0079}$	$0.095^{+0.013}_{-0.010}$
pomeron coupling (mb)	$18.4^{+1.8}_{-2.0}$	$18.8^{+1.7}_{-2.0}$	$18.5^{+2.1}_{-2.6}$
ρ/ω intercept−1	$-0.535^{+0.051}_{-0.059}$	$-0.518^{+0.050}_{-0.058}$	$-0.540^{+0.059}_{-0.067}$
ρ/ω coupling $C_-(pp)$ (mb)	$-31.6^{+7.6}_{-11.5}$	$-28.9^{+6.8}_{-10.4}$	$-32.5^{+8.8}_{-13.9}$
a/f intercept−1	$-0.338^{+0.054}_{-0.055}$	$-0.355^{+0.056}_{-0.057}$	$-0.346^{+0.067}_{-0.066}$
a/f coupling (mb)	$58.8^{+8.7}_{-6.8}$	$61.5^{+9.8}_{-7.7}$	$60.4^{+10.5}_{-7.9}$
σ_{tot}(1.8 TeV) (mb)	$77.3^{+2.6}_{-2.7}$	77.2 ± 2.6	$77.2^{+3.6}_{-3.3}$
σ_{tot}(10 TeV) (mb)	$106.8^{+7.1}_{-6.3}$	$106.3^{+6.8}_{-6.2}$	$106.5^{+9.7}_{-7.8}$
σ_{tot}(14 TeV) (mb)	$113.9^{+8.3}_{-7.4}$	$113.2^{+8.0}_{-7.1}$	$113.5^{+11.4}_{-9.0}$

Table 5: Simple pole fit to total pp and $\bar{p}p$ cross sections, and to the ρ parameter, with non-degenerate $C = +1$ and $C = -1$ meson exchanges, and using the alternative data set of Ref. [8].

Our best estimate for the pomeron intercept is then:

$$\alpha_P = 1.0964^{+0.0115}_{-0.0091} \tag{4}$$

based on the 2σ-filtered PDG data, in the non-degenerate case.

At this point, the only additional piece of data might be the direct observation of the pomeron, *i.e.* of a 2^{++} glueball. Using the observed X(1900) mass as confirmed by the WA91 collaboration [14], 1918 ± 12 MeV, for a $I^G J^{PC} = 0^+2^{++}$ state, $f_2(1900)$, and using $\alpha' = 0.250$ GeV^{-2} [5], we obtain $\alpha_P = 1.0803 \pm 0.012$. This is the value of the intercept for 1-pomeron exchange. The intercepts that we obtained in Tables 2, 3, 4 and 5 from scattering data cannot be directly compared with this value, as they include the effect of multiple exchanges, of pomerons and reggeons. But the values we have derived are certainly compatible with the WA91 measurement. Note however that an intercept of 1.094 would be in perfect agreement with the WA91 observation for a slope $\alpha' = 0.246$ GeV^{-2}. Hence it would be dangerous to mix this piece of information with the t-channel information.

Finally, we can place constraints on physics beyond one-pomeron exchange. Using 2σ-filtered data, we obtain an upper bound on the ratio of the 2-pomeron coupling to that of the pomeron to be 4.7% at 70% C.L., and including such a contribution would bring the best value for the intercept of the 1-pomeron exchange term to $1.126^{+0.051}_{-0.082}$. Also at the 70% C.L., the ratio of the coupling of an odderon to that of a pomeron is smaller than 0.1% (the best odderon intercept would then be 1.105 and the pomeron intercept become 1.099). This would correspond to 0.08 mb at the Tevatron. As for the "hard pomeron", there is no trace of it in the data. Constraining its intercept to being larger than 1.3 leads to an upper bound on the ratio of its coupling to that of the pomeron of 0.9% (the soft pomeron intercept then becomes 1.065). This would correspond to a maximum hard contribution of 19 mb at the Tevatron. Consequently, the simple pole model fits to total cross sections are very successful.

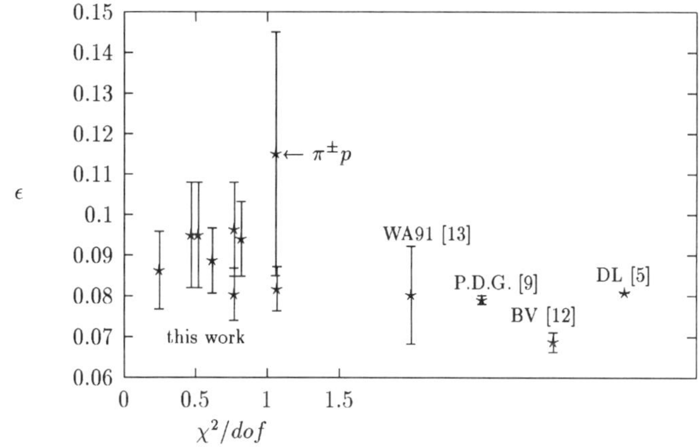

Figure 4: Our results for the pomeron intercept, compared with others in the literature. The values of the $\chi^2/d.o.f.$ are indicated for the points of this work only.

We show in Fig. 4 the results obtained in this paper together with other estimates present in the literature. All the points from this work have an acceptable χ^2, and the main difference between them is either the filtering of data or the physics of lower trajectories. Since all these estimates are acceptable, we conclude that the pomeron intercepts as high as 1.11, and as low as 1.07, are possible. When comparing with other works in the literature, we have explained that the use of a small energy cutoff leads to smaller intercepts, and reflects the fact that sub-leading meson trajectories are to be included. Note however that the original DL fit [6] used the same cutoff as ours, but used a different definition of χ^2 [15].

Our errors are much larger than those of other estimates because we fully take into account the correlation of the various parameters, and because our statistical analysis of the data is much more careful than previous ones. Though these results depend only on pp and $\bar{p}p$ data, we have argued that little could be learned from other hadronic reactions, given that they are measured at low energy. In particular, we point out that our fit to total cross sections, as shown in Fig. 3, is indistinguishable from the DL fit for $\sqrt{s} < 300$ GeV, and hence the parametrisation we propose is expected to fit well the total γp cross sections, as well as the πp and Kp data.

References

[1] I. Ya Pomeranchuk, *Sov. Phys. JETP* **34(7)** (1958) 499; See *e.g.* P.D.B. Collins, An Introduction to Regge Theory and High-Energy Physics, (Cambridge University Press, Cambridge: 1977).

[2] ZEUS Collaboration (M. Derrick et al.), *Phys. Lett.* **B315** (1993), 481; H1 Collaboration (T. Ahmed et al.), *Nucl. Phys.* **B429** (1994) 477.

[3] J.D. Bjorken, *Phys. Rev.* **D47** (1992) 101.

[4] Available at http://www-pdg.lbl.gov/xsect/contents.html, courtesy of A. Baldini, V. Flaminio, W.G. Moorhead, and D.R.O. Morrison, CERN; and COMPAS Group, IHEP, Protvino, Russia.

[5] A. Donnachie and P.V. Landshoff, *Nucl. Phys.* **B244** (1984) 322; *Nucl. Phys.* **B231** (1984) 189; *Nucl. Phys.* **B267** (1986) 690.

[6] A. Donnachie and P.V. Landshoff, *Phys. Lett.* **B296** (1992) 227, e-Print Archive: hep-ph/9209205.

[7] UA4.2 Collaboration, C. Augier et al., *Phys. Lett.* **B315** (1993) 503.

[8] S. Hadjitheodoridis, Ph.D. Thesis, Brown University (May 1989).

[9] see e.g. K. Kang and S.K. Kim, preprint BROWN-HET-1008 (Oct. 1995), talk given at 7th Rencontres de Blois: *Frontiers in Strong Interactions - 6th International Conference on Elastic and Diffractive Scattering*, Blois, France, 20-24 June 1995, e-Print Archive: hep-ph/9510438; K. Kang, P. Valin and A.R. White, *Nuov. Cim.* **107A** (1994) 2103.

[10] Review of Particle Physics, Particle Data Group, R.M. Barnett et al., *Phys. Rev.* **D54** (1996).

[11] F. James, M. Roos *Computer Physics Commun.* **10** (1975) 343; F. James and M. Goosens, Minuit Reference Manual, CERN Program Library Long Writeup D506 (March 1994), available from ftp://asis01.cern.ch/cernlib/doc/ps.dir/minuit.ps.

[12] Numerical Recipes in C, William H. Press et al. (Cambridge University Press, Cambridge: 1988), Chapter 14.

[13] A. Bueno and J. Velasco, preprint IFIC-96-15 (May 1996).

[14] WA91 Collaboration, F. Antinori et al., *Phys. Lett.* **B353** (1995) 589, WA76 Collaboration, T.A. Armstrong et al., *Phys. Lett.* **B228** (1989) 536.

[15] P.V. Landshoff, private communication.

ARE WE REALLY MEASURING ρ, THE PHASE OF THE NUCLEAR SCATTERING AMPLITUDE?

M. M. BLOCK

Department of Physics and Astronomy, Northwestern University, Evanston, IL 60208, U.S.A.

The interference at small $|t|$ of the Coulomb scattering amplitude f_c and the nuclear amplitude f_n is used to measure the phase of the nuclear scattering amplitude, and hence the ρ-value, where $\rho \equiv \left(\frac{\Re e f_n}{\Im m f_n}\right)_{t=0}$. The "normal" analysis of $\bar{p}p$ and pp elastic scattering uses a 'spinless' Coulomb amplitude, *i.e.*, a Rutherford amplitude $(2\sqrt{\pi}\alpha/t)$ multiplied by a Coulomb form factor $G^2(t)$, an *ansatz* that pretends that the nucleon does not have any magnetic scattering. In this note, we investigate the role of the anomalous magnetic moment of the nucleon, $\kappa \approx 1.79$. Given the method of analysis currently used by most published experiments, we conclude that the current experimentally inferred values of ρ for $\bar{p}p$ should be *systematically lowered* by ≈ 0.005—0.0100 and, correspondingly, the ρ values for pp should be *systematically raised* by the same amount. We discuss the theoretical uncertainties and a method of experimentally minimizing them.

1 Introduction

This paper is dedicated to Peter Carruthers, an enthusiastic and eclectic physicist, a fabulous fisherman and a very dear friend.

We will only calculate electromagnetic amplitudes accurate to order α, *i.e.*, the one-photon exchange diagram shown in Fig. 1. Further, we will consider only high energy scattering ($E_{\rm lab} \gg m$, where m is the nucleon mass) in the region of small $|t|$, where t is the squared 4-momentum transfer. We will measure m and $E_{\rm lab}$ in GeV and t in $({\rm GeV})^2$, and will use $\hbar = c = 1$.

1.1 'Spinless' Coulomb Scattering

If we consider 'spinless' proton-antiproton Coulomb scattering, the relevant Feynman diagram is shown in Fig. 1, with $V^\mu = G\,(p_i + p_f)^\mu$ and $G(t)$ is the electromagnetic charge form factor of the nucleon.

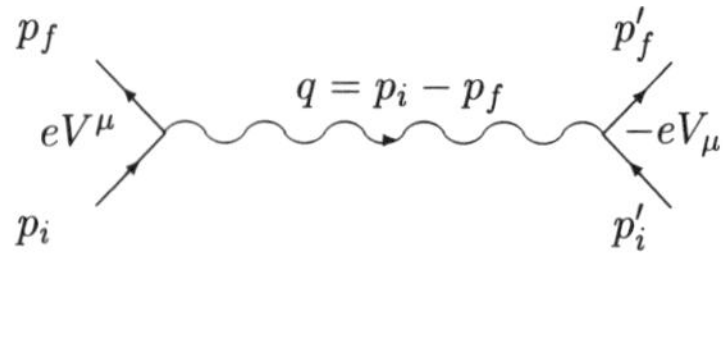

Figure 1: One-Photon Feynman Diagram for Proton—Anti- Proton Coulomb Scattering, $p_i + p'_i \to p_f + p'_f$, with Couplings eV^μ and $-eV_\mu$.

The electromagnetic differential cross section is readily evaluated as

$$\frac{d\sigma}{dt} = \pi \left| \frac{\mp 2G^2(t)\alpha}{\beta_{\text{lab}}|t|} \times \left(1 - \frac{|t|}{4mE_{\text{lab}}}\right) \right|^2, \tag{1}$$

where the upper (lower) sign is for like (unlike) charges, t is the (negative) 4-momentum transfer squared, and m is the nucleon mass.

For small angle scattering and at high energies, the correction term $\frac{|t|}{2mE_{\text{lab}}}$ becomes negligible and $\beta_{\text{lab}} \to 1$, so eq. (1) goes over into the well-known Rutherford scattering formula,

$$\frac{d\sigma}{dt} = \pi \left| \frac{\mp 2\alpha G^2(t)}{|t|} \right|^2, \tag{2}$$

where the electromagnetic charge form factor $G(t)$ is commonly parameterized by the dipole form

$$G(t) = \frac{1}{\left(1 - \frac{t}{\Lambda^2}\right)^2}, \tag{3}$$

where $\Lambda^2 = 0.71$, if t is measured in $(\text{GeV})^2$. We note that this is the Coulomb amplitude that is normally used in the analysis of $\bar{p}p$ and pp elastic scattering, *i.e.*, the 'spinless' analysis[1].

1.2 $\bar{p}p$ Scattering, Including Magnetic Scattering

The relevant Feynman diagram is shown in Fig. 1, where magnetic scattering is explicitly taken into account via the anomalous magnetic moment κ (≈ 1.79).

The fundamental electromagnetic interaction is

$$eV^\mu = e\left(F_1\gamma^\mu + i\frac{\kappa}{2m}F_2\sigma^{\nu\mu}q_\nu\right), \qquad q = p_f - p_i \tag{4}$$

which has two form factors $F_1(q^2)$ and $F_2(q^2)$, both normalized to 1 at $q^2 = 0$. The anomalous magnetic moment of the nucleons is κ and m is the nucleon mass. Because of the rapid form factor dependence on t, the annihilation diagram for $\bar{p}p$ scattering (or the exchange diagram for pp scattering) is negligible in the small $|t|$ region of interest and has been ignored. The interaction of eq. (4) is most simply treated by using Gordon decomposition and can be rewritten as $eV^\mu = e\left[(F_1 + \kappa F_2)\gamma^\mu - \kappa F_2\left(\frac{p_f+p_i}{2m}\right)^\mu\right]$. Thus, using this form, the matrix element for the scattering is given by

$$\begin{aligned} M = & \mp e\bar{u}(p_f)\left[-\kappa F_2\left(\frac{p_f + p_i}{2m}\right)^\mu + (F_1 + \kappa F_2)\gamma^\mu\right]u(p_i) \times \frac{1}{t} \times \\ & e\bar{u}(p'_f)\left[-\kappa F_2\left(\frac{p'_f + p'_i}{2m'}\right)_\mu + (F_1 + \kappa F_2)\gamma_\mu\right]u(p'_i), \end{aligned} \tag{5}$$

where the upper (lower) sign is for $\bar{p}p$ (pp) scattering. A straightforward, albeit laborious calculation, gives a differential scattering cross section

$$\begin{aligned}\frac{d\sigma}{dt} &= 4\pi\frac{\alpha^2}{\beta_{\rm lab}^2 t^2}\times \\ &\left\{(F_1+\kappa F_2)^4\left[1+\frac{t}{2}\left(\frac{1}{mE_{\rm lab}}+\frac{1}{E_{\rm lab}^2}\right)+\frac{t^2}{8m^2E_{\rm lab}^2}\right]\right. \\ &-2(F_1+\kappa F_2)^2\left[\kappa^2F_2^2\left(1+\frac{t}{4m^2}\right)+2\kappa F_1F_2\right]\times \\ &\qquad\left[1+\frac{t}{2}\left(\frac{1}{mE_{\rm lab}}+\frac{1}{2E_{\rm lab}^2}\right)\right] \\ &\left.+\left[\kappa^2F_2^2\left(1+\frac{t}{4m^2}\right)+2\kappa F_1F_2\right]^2\left[1+\frac{t}{2mE_{\rm lab}}+\frac{t^2}{16m^2E_{\rm lab}^2}\right]\right\}. \qquad (6)\end{aligned}$$

We now introduce the *electric* and *magnetic* form factors, $G_E(t)$ and $G_M(t)$, defined as

$$G_E(t) \equiv F_1(t)+\frac{\kappa t}{4m^2}F_2(t) \qquad \text{and} \quad G_M(t)\equiv F_1(t)+\kappa F_2(t), \qquad (7)$$

and rewrite the differential cross section of eq. (6) as

$$\begin{aligned}\frac{d\sigma}{dt} &= 4\pi\frac{\alpha^2}{\beta_{\rm lab}^2t^2}\times \\ &\left\{\left(\frac{G_E^2(t)-\frac{t}{4m^2}G_M^2(t)}{1-\frac{t}{4m^2}}\right)^2\left(1+\frac{t}{2mE_{\rm lab}}\right)+G_M^2(t)\frac{G_E^2(t)-\frac{t}{4m^2}G_M^2(t)}{1-\frac{t}{4m^2}}\frac{t}{2E_{\rm lab}^2}\right. \\ &\left.+\left[G_M^4(t)+\frac{1}{2}\left(\frac{G_E^2(t)-G_M^2(t)}{1-\frac{t}{4m^2}}\right)^2\right]\frac{t^2}{8m^2E_{\rm lab}^2}\right\}. \qquad (8)\end{aligned}$$

We can parameterize these new form factors with

$$\begin{aligned}G_E(t) &= G(t) &= \frac{1}{\left(1-\frac{t}{\Lambda^2}\right)^2} \qquad \text{where} \quad \Lambda^2=0.71, \\ G_M(t) &= (1+\kappa)G(t) &= \frac{1+\kappa}{\left(1-\frac{t}{\Lambda^2}\right)^2},\end{aligned} \qquad (9)$$

with t in GeV/c^2, and where $G(t)$ is the dipole form factor already defined in eq. (3), *i.e.*, the form factor that is traditionally used in experimental analyses[1].

We now expand eq. (8) for *very small* $|t|$, and find that

$$\frac{d\sigma}{dt} \approx 4\pi\frac{\alpha^2}{\beta_{\rm lab}^2t^2}G^4(t)\left\{1-\kappa(\kappa+2)\frac{t}{2m^2}+\frac{t}{2mE_{\rm lab}}+(\kappa+1)^2\frac{t}{2E_{\rm lab}^2}\right\}, \qquad (10)$$

where the *new term* in t, compared to eq. (1), is $-\frac{\kappa(2+\kappa)}{2m^2}t+\kappa(\kappa+1)\frac{t}{2E_{\rm lab}^2} \approx 1+3.86|t|-\frac{3.39}{E_{\rm lab}^2}|t|$, and is due to the anomalous magnetic moment of the proton (antiproton). To get an estimate of its effect, we note that $G^4(t) \approx 1-11.26|t|$, in our units where t is in GeV^2. We note that the new term is *not negligible* in comparison to the squared form factor, reducing the form factor effect by about 35% if the energy $E_{\rm lab}$ is large compared to m. In this limit, we find now a t-dependent term, independent of the energy $E_{\rm lab}$, *i.e.*,

$$\frac{d\sigma}{dt} \approx 4\pi\frac{\alpha^2}{t^2}G^4(t)\left\{1+3.86|t|\right\}, \tag{11}$$

which is to be compared with the 'spinless' Rutherford formula of eq. (2).

2 Effects on Experimental Analysis of Elastic Scattering

UA4/2 has recently made a precision measurement[2] of $\bar{p}$-p scattering at $\sqrt{s} = 541$ GeV, at the S$\bar{p}p$S at CERN, in order to extract the ρ value for elastic scattering, with $\rho \equiv \left(\frac{\Re e f_n}{\Im m f_n(}\right)_{t=0}$, where $f_n(t=0)$ is the forward *nuclear* scattering amplitude. They constrain the total cross section by an independent measurement[3] of $(1+\rho^2)\sigma_{\rm tot} = 63.3 \pm 1.5$ mb. For their published ρ-value of 0.135 ± 0.015, this implies that they fix the total cross section at $\sigma_{\rm tot} = 62.17 \pm 1.5$ mb. We now reanalyze this experiment, taking into account the magnetic scattering.

2.1 Spinless Analysis Neglecting Magnetic Scattering

The experimenters parameterized the nuclear slope amplitude, as usual, as

$$f_n(|t|) = \frac{\sigma_{\rm tot}(\rho+i)}{4\sqrt{\pi}}e^{-b|t|/2}, \tag{12}$$

and measured the nuclear slope parameter $b = 15.5 \pm 0.2$ GeV^{-2}. They fit the $\bar{p}p$ elastic scattering data at $\sqrt{s} = 541$ GeV over the t-interval $0.00075 \le |t| \le 0.12$ GeV^2. They used for the Coulomb amplitude the 'spinless' Rutherford amplitude, modified by a Coulomb phase factor $i\alpha\phi(t)$, *i.e.*,

$$f_c(t) = \frac{2\sqrt{\pi}\alpha}{|t|}G^2(t)e^{i\alpha\phi(t)}, \tag{13}$$

where the phase[4] $\phi(t)$ is given by

$$\phi(t) = \mp\left\{\gamma + \ln\left(\frac{b|t|}{2}\right) + \ln\left(1+\frac{8}{b\Lambda^2}\right) + \left(\frac{4|t|}{\Lambda^2}\right)\ln\left(\frac{4|t|}{\Lambda^2}\right) + \frac{2|t|}{\Lambda^2}\right\}, \tag{14}$$

where $\gamma = 0.577\ldots$ is Euler's constant, b is the slope parameter, $\Lambda^2[= 0.71$ GeV$^2]$ appears in the dipole fit to the proton's electromagnetic form factor, $G(t)$. The upper sign is for pp and the lower sign for $\bar{p}p$. Using these 'standard' parameterizations[1], the differential elastic scattering cross section is

$$\frac{d\sigma}{d|t|} = |f_c+f_n|^2 = \frac{4\pi\alpha^2}{t^2}G^4(t) + \frac{\alpha\sigma_{\rm tot}}{|t|}(\rho+\alpha\phi(t))G^2(t)e^{-b|t|/2} + \frac{\sigma_{\rm tot}^2(1+\rho^2)}{16\pi}e^{-b|t|}. \tag{15}$$

We now introduce the parameter t_0, defined as the absolute value of t where the nuclear and Coulomb amplitudes have the same magnitude, *i.e.*, $t_0 = \frac{8\pi\alpha}{\sigma_{\rm tot}} = \frac{1}{14.00\sigma_{\rm tot}}$, when $\sigma_{\rm tot}$ is in mb, and t_0 is in GeV^2. For $\sigma_{\rm tot} = 62.17$ mb, we find that $t_0 = 0.00115$ GeV^2. We can now rewrite the differential cross section, in the region $t_{\rm min} \le t \le t_{\rm max}$, as

$$\frac{d\sigma}{d|t|} = \frac{\sigma_{\rm tot}^2}{16\pi}\left\{G^4(t)\frac{t_0^2}{t^2} + 2\frac{t_0}{|t|}(\rho + \alpha\phi(t))G^2(t)e^{-b|t|/2} + (1+\rho^2)e^{-b|t|}\right\}, \tag{16}$$

the formula used by the UA4/2 group to analyze their experimental data. They extracted the value $\rho = 0.135 \pm 0.015$, with a statistical error of 0.007. We emphasize that their analysis, using eq. (16), neglected the effects of the anomalous magnetic moments of the nucleons.

2.2 *'Spinless' Analysis, Taking into Account the Magnetic Scattering*

For our small $|t|$ analysis, we approximate $G^4(t)$ as $G^4(t) \approx 1 - 2a|t|$, where $a \equiv \frac{4}{\Lambda^2} = 5.6338$. We further write $G^2(t)e^{-b|t|/2} \approx 1 - (a + \frac{b}{2})|t|$. However, if we take into account the anomalous magnetic moments, we see from eq. (11) that we could have written $f_c(t) \approx 2\sqrt{\pi}\frac{\alpha}{|t|}G^2(t)\left\{1 + 1.93|t|\right\}e^{i\alpha\phi(t)}$, an energy-independent result. Literally, we have used the *'spinless' ansatz* that the Coulomb amplitude is given by the *square root* of the Coulomb cross section. Expanding the above equation in $|t|$, incorporating the new factor of $1 + 1.93|t|$, we can rewrite in a concise form the 'correct' Coulomb amplitude as

$$f_c(t) \approx 2\sqrt{\pi}\frac{\alpha}{|t|}G_{\rm eff}^2(t)e^{i\alpha\phi(t)}, \tag{17}$$

where $G_{\rm eff}^2(t) \approx 1 - a'|t| = 1 - (a - 1.93)|t| = 1 - 3.674|t|$. Thus, $a' = 3.674$, and we can now write a modified form for the cross section which mimics eq. (16) as

$$\frac{d\sigma'}{d|t|} = \frac{\sigma_{\rm tot}^2}{16\pi}\left\{G_{\rm eff}^4(t)\frac{t_0^2}{t^2} + 2\frac{t_0}{|t|}(\rho' + \alpha\phi(t))G_{\rm eff}^2(t)e^{-b|t|/2} + (1+\rho'^2)e^{-b|t|}\right\}, \tag{18}$$

for $t_{\rm min} \le |t| \le t_{\rm max}$, by using $G_{\rm eff}(t)$ in place of $G(t)$, *i.e.*, by replacing a by a' and ρ by ρ'.

In order to extract the new value of ρ' from eq. (18) without having to refit directly the experimental data, we will require that the *integral* of eq. (18) be equal to the *integral* of eq. (16), from $t_{\rm min}$ to $t_{\rm max}$. This insures that we fit the *measured events* with both formulae, allowing us to solve for ρ', given the UA4/2 published value of $\rho = 0.135$. Thus, we require

$$\int_{t_{\rm min}}^{t_{\rm max}}\frac{d\sigma'}{d|t|} = \int_{t_{\rm min}}^{t_{\rm max}}\frac{d\sigma}{d|t|}. \tag{19}$$

To obtain an approximate analytical solution, we can set equal the contributions of the terms $(1+\rho'^2)e^{-b|t|}$ and $(1+\rho^2)e^{-b|t|}$, and expand in $|t|$ to first order. We find the approximate solution for ρ', the *corrected* value of ρ, valid when we set the contribution of $\alpha\phi(t) = 0$,

$$\rho' \approx \rho + (a' - a)\left(t_0 + \rho\frac{t_{\rm max} - t_{\rm min}}{\ln(t_{\rm max}/t_{\rm min})}\right) \qquad \text{if} \quad t_0 \ll \rho\frac{t_{\rm max} - t_{\rm min}}{\ln(t_{\rm max}/t_{\rm min})}. \tag{20}$$

Neglecting the $\alpha\phi(t)$ contribution is a reasonable approximation, since the average value of $\alpha\phi(t) = -0.00036 \approx 0$ over the t-range in question. We find, using the UA4/2 extracted values described above, that $\Delta\rho = \rho' - \rho = \rho' - 0.135 = -0.0114$. An *exact numerical solution* of eq. (19), where we expand to first order in $|t|$, yields the final answer $\Delta\rho_{\mathrm{UA4}} = -0.0105$. This shift in ρ is *comparable* to the quoted total error of ± 0.015 and is *larger* than the statistical error of ± 0.007. The solution is completely insensitive to the slope parameter b, as well as to $\alpha\phi(t)$. It mainly depends on the value of ρ that's measured, and critically on the t interval that was measured. We note that these conclusions are in contrast with the analysis of this problem made by N. Buttimore[6], which did not take into account the experimental t interval.

Since all high energy ρ values come from experiments that measure about the same range of t and similar b values, it is likely that *all* the ρ values for $\bar{p}p$ scattering need to be *lowered* by ~ 0.005—0.010 and that all pp values need to be *raised* by about he same amount. For example, E710 measured[5] $\rho_{\bar{p}p} = 0.134 \pm 0.069$, $\sigma_{\mathrm{tot}} = 72.2 \pm 2.7$ mb and $b = 16.72\ \mathrm{GeV}^{-2}$, with $t_{\min} = 0.00075\ \mathrm{GeV}^2$ and $t_{\max} = 0.077\ \mathrm{GeV}^2$, at $\sqrt{s} = 1.8$ TeV. In this case, the exact numerical solution of eq. (19) gives $\Delta\rho_{\mathrm{E710}} = -0.004$. Thus, any global fit to the ρ values of $\bar{p}p$ and pp scattering should take these *systematic shifts* in the ρ values (at *all* energies) into account.

2.3 A Nuclear Model Taking Account of Magnetic Scattering

In the preceding Section, we emulated the 'spinless' analysis by *ignoring* the structure of the nuclear scattering, and *not distinguishing* between spin flip and non-spin flip of the nuclear scattering, which we will now remedy. In this Section, we will introduce a toy nuclear model, which mimics the known electromagnetic matrix element of eq. (5), and adds *coherently* with it. We thus write the total matrix element for coherent nuclear and Coulomb scattering, taking into account the Bethe phase as expressed by Cahn[4], as

$$\begin{aligned}
M_t' &= \pm\frac{4\pi\alpha e^{i\alpha\phi(t)}}{|t|}\bar{u}(p_f)\left[-\kappa F_2\left(\frac{p_f+p_i}{2m}\right)^{\mu} + (F_1+\kappa F_2)\gamma^{\mu}\right]u(p_i)\times \\
&\quad \bar{u}(p_f')\left[-\kappa F_2\left(\frac{p_f'+p_i'}{2m'}\right)_{\mu} + (F_1+\kappa F_2')\gamma_{\mu}\right]u(p_i') \quad + \\
&\quad \frac{4\pi g}{m^2}\bar{u}(p_f)\left[-\kappa_N H_2\left(\frac{p_f+p_i}{2m}\right)^{\mu} + (H_1+\kappa_N H_2)\gamma^{\mu}\right]u(p_i)\times \\
&\quad \bar{u}(p_f')\left[-\kappa_N H_2\left(\frac{p_f'+p_i'}{2m}\right)_{\mu} + (H_1+\kappa_N H_2)\gamma_{\mu}\right]u(p_i'), \qquad (21)
\end{aligned}$$

where now the upper sign is for $\bar{p}p$ and the lower sign is for pp scattering, since we introduced $|t|$ into eq. (21). We have substituted $e^2 = 4\pi\alpha$ and defined the strong coupling analog of α as $h^2 = 4\pi g$, where h is the (complex) nuclear 'charge', κ_N is the nuclear 'anomalous magnetic moment' and $H_1(t)$ and $H_2(t)$ are the nuclear form factors. We have replaced the electromagnetic propagator $|t|$ by the nuclear propagator m^2. Later, we will use the optical theorem to fix the real and imaginary portions of h^2 and will also introduce "electric" and "magnetic" nuclear form factors G_{E_N} and G_{M_N}, analogous to eq. (7), with

$$G_{E_N}(t) \equiv H_1(t) + \frac{\kappa_N t}{4m^2} H_2(t) \qquad \text{and} \quad G_{M_N}(t) \equiv H_1(t) + \kappa_N H_2(t), \tag{22}$$

and

$$\begin{aligned} G_{E_N}(t) &= G_N(t) &&= e^{bt/4} \\ G_{M_N}(t) &= (1+\kappa_N) G_N(t) &&= (1+\kappa_N) e^{bt/4}. \end{aligned} \tag{23}$$

The squaring of M_t' in eq. (21) will give rise to three terms,

$$|M_t|^2 = |M_c|^2 \pm 2|M_c|(\Re e M_N + \alpha\phi) + |M_N|^2, \tag{24}$$

since $e^{i\alpha\phi(t)} \approx 1 + i\alpha\phi(t)$. The first term of eq. (24) corresponds to pure Coulomb scattering, the last term to pure nuclear scattering and the term $2|M_c|(\Re e M_N + \alpha\phi)$ to the coherent interference cross section between nuclear and Coulomb amplitudes. We note that the Dirac structure of *all three terms* in eq. (24) is the *same*, which greatly simplifies the evaluation. We have already calculated the term $|M_c|^2$ and the substitution of $\alpha \to g$, $F_1 \to H_1$, $F_2 \to H_2$ and $\frac{1}{|t|} \to \frac{1}{m^2}$ in the Coulomb term gives us the nuclear term $|M_N|^2$. Thus we find by inspection of eq. (10) at $t = 0$, using the optical theorem, that

$$\left(\frac{d\sigma_N}{dt}\right)_{t=0} = \frac{\sigma_{\rm tot}^2(1+\rho^2)}{16\pi} = 4\pi \frac{g^2}{\beta_{\rm lab}^2 m^4}, \tag{25}$$

since $G_N(0) \equiv 1$. Inspection of eq. (25) yields the equivalent statement that the (complex) value of the nuclear coupling is $\frac{g}{\beta_{\rm lab} m^2} = (\rho + i)\frac{\sigma_{\rm tot}}{8\pi}$. Again, by inspection of eq. (10), using eq. (25), we can now write the nuclear differential cross section at small $|t|$ and large $E_{\rm lab}$ as

$$\frac{d\sigma_N}{dt} = \frac{\sigma_{\rm tot}^2(1+\rho^2)}{16\pi} e^{bt} \left\{1 - \frac{\kappa_N(\kappa_N+2)}{2m^2} t\right\}. \tag{26}$$

The linear term in t in the brackets of eq. (26) is clearly the spin flip term induced by the nuclear "anomalous magnetic moment" κ_N and goes to zero in the forward direction, as must be true for spin flip amplitudes.

Again, with a lengthy, but straightforward calculation, we find that the interference cross section $\frac{d\sigma_{\rm CN}}{dt}$, for small $|t|$ and large $E_{\rm lab}$, is

$$\frac{d\sigma_{\rm CN}}{dt} = \mp \frac{\alpha G^2(t)}{\beta_{\rm lab}|t|} \sigma_{\rm tot} e^{bt/2} \left[\rho + \alpha\phi(t)\right] \left[1 - \frac{\kappa + \kappa_N + \kappa\kappa_N}{2m^2} t\right]. \tag{27}$$

Introducing again the parameter $t_0 = \frac{8\pi\alpha}{\sigma_{\rm tot}}$, we write, in the small $|t|$ and large $E_{\rm lab}$ limit, the elastic differential scattering cross section $\frac{d\sigma}{d|t|}$ for *coherent* Coulomb and nuclear interactions as

$$\frac{d\sigma}{d|t|} = \frac{\sigma_{\rm tot}^2}{16\pi}\left\{\frac{t_0^2}{t^2}G^4(t)\left[1+\frac{\kappa(\kappa+2)}{2m^2}|t|\right]\right.$$
$$\mp 2G^2(t)\frac{t_0}{|t|}\left(\rho+\alpha\phi(t)\right)e^{-b|t|/2}\left[1+\frac{\kappa+\kappa_N+\kappa\kappa_N}{2m^2}|t|\right]$$
$$\left.+(1+\rho^2)e^{-b|t|}\left[1+\frac{\kappa_N(\kappa_N+2)}{2m^2}|t|\right]\right\}. \tag{28}$$

In order to use eq. (28) for the small $|t|$ region used in experiments, we first expand the exponential of the nuclear term for small $|t|$, *i.e.*,

$$\left(\frac{d\sigma}{d|t|}\right)_N \approx \frac{\sigma_{\rm tot}{}^2}{16\pi}(1+\rho^2)e^{-b'|t|}, \qquad \text{where } b' = b - \frac{\kappa_N(\kappa_N+2)}{2m^2}. \tag{29}$$

Effectively, it is b' that is measured, *not* b. Thus we finally write eq. (28) as

$$\frac{d\sigma}{d|t|} \approx \frac{\sigma_{\rm tot}^2}{16\pi}\left\{\frac{t_0^2}{t^2}G^4(t)\left[1+\frac{\kappa(\kappa+2)}{2m^2}|t|\right]\right.$$
$$\mp 2G^2(t)\frac{t_0}{|t|}\left(\rho+\alpha\phi(t)\right)e^{-b'|t|/2}\left[1+\frac{2\kappa+2\kappa\kappa_N-\kappa_N^2}{4m^2}|t|\right]$$
$$\left.+(1+\rho^2)e^{-b'|t|}\right\}. \tag{30}$$

We really have no deep knowledge of a value that is appropriate for κ_N, let alone its sign. Since the known polarization at Fermilab energies[7,8] is small, of the order of several percent, a magnitude compatible with it being produced by Coulomb interactions, this suggests that $|\kappa_N| \lesssim \kappa$. Interestingly, if we set $\kappa = \kappa_N$, we find the *same* interference term of eq. (18) where we "ignored" spin, *i.e.*, $2G^2(t)\frac{t_0}{|t|}e^{-b|t|/2}(1+1.93|t|)$. Thus, the numerical evaluation made earlier in Section 2.1 is valid—namely, for UA4/2, the ρ value changes by ≈ -0.01.

On the other hand, even if we set $\kappa_N = 0$, we *still* would have an interference term proportional to $(1+1.02|t|)$. We see from eq. (20) that for $\kappa_N = 0$ that the shift in ρ, which is proportional to $a' - a$, is now reduced by $1.02/1.93 = 0.53$. Thus, the ρ value for $\kappa_N = 0$ is shifted by ≈ -0.006. We conclude that for reasonable values of the parameter κ_N, there is a *significant* shift in the ρ value of UA4/2 due to the anomalous magnetic moment of the nucleon. The *theoretical uncertainty* in the ρ-value can be reduced if the experimenters break their data up into *two* distinct regions—Region 1 being the interference region from $t_{\rm min}$ to $\approx 10\times t_0$ and Region 2 from $10\times t_0$ to $t_{\rm max}$. In the case of UA4/2, this would change the uncertainty in $\Delta\rho_{\rm UA4}$ from ≈ 0.006 to ≈ 0.001, if we found ρ from Region 1 and $\sigma_{\rm tot}$ and b from Region 2, if $\kappa_N = 0$.

The other theoretical uncertainty appears in the value of the phase $\alpha\phi(t)$. The original derivation[4] assumed that we had 'spinless' scattering, that the nuclear amplitude was $e^{b't/2}$, where b' was the measured slope, and that the Coulomb amplitude was $\frac{\alpha}{t}G^2(t)$. We see that we really should be using $\frac{\alpha}{t}G^2_{\rm eff}(t)$, or, in other words, using $\Lambda'^2 \approx 1.0$ rather than

$\Lambda^2 = 0.71$, in eq. (14). Fortunately, this is a *very small* change, and increases the $\bar{p}p$ ρ-value by ≈ 0.001, thus contributing negligibly to $\Delta\rho_{\mathrm{UA4}}$.

In conclusion, it seems sensible for experimenters to redo their data analysis using $\kappa_N = \kappa$, *i.e.*, using $G_{\mathrm{eff}}(t)$ and Λ'^2 rather than $G(t)$ and Λ^2, in the two distinct t-regions described above. This procedure allows the experimenter to control the theoretical uncertainties, *i.e.*, our lack of knowledge of the nuclear amplitudes.

Acknowledgments

This work was partially supported by Department of Energy grant DA-AC02-76-Er02289 Task B.

References

1. M. M. Block and R. N. Cahn, Rev. Mod. Phys. **57**, 563 (1985).
2. UA4 Collaboration, C. Augier *et al.*, CERN-PPE-93-115 July (1993); International Conference on Elastic and Diffractive Scattering, Proceedings of the Vth Blois Workshop, Brown University, World Scientific, Editors, H. Fried, K. Kang and C.-I. Tan, p. 7, June (1993); Phys. Lett. **B316** , 448 (1993).
3. UA4 Collaboration, M. Bozzo *et al.*, Phys. Lett. **B147** , 392 (1984).
4. R. N. Cahn, Zeitschr. für Phys. **C15**, 253 (1982).
5. E710 Collaboration, N. A. Amos *et al.*, Phys. Rev. Lett. **68**, 2433 (1992); International Conference on Elastic and Diffractive Scattering, Proceedings of the Vth Blois Workshop, Brown University, World Scientific, Editors, H. Fried, K. Kang and C.-I. Tan, p. 59, June (1993).
6. N. H. Buttimore,"EM–strong interactions in proton proton elastic scattering", Proceedings of the Adriatico Research Conference on Polarization Dynamics in Nuclear and Particle Physics, ICTP, Trieste, Italy, World Scientific, Editors, A. O. Barut, N. Paver, A. Penzo and R. Raczka, p. 228, (1993).
7. R. V. Kline, "High Energy Physics with Polarized Beams and Targets", Proceedings of the Argonne Symposium, 1976 Edited by M. L. Marshak(AIP, New York, 1977), p. 152;
W. de Boer *et al.*, Phys. Rev. Lett. **34**, 558 (1975).
8. N. H. Buttimore, E. Gotsman and E. Leader, Phys. Rev. **D 18**, 694 (1978).

Diffractive Production at Collider Energies: Soft Diffraction and Dino's Paradox

Chung-I Tan
Department of Physics, Brown University
Providence, RI 02912

Abstract

One of the more interesting developments from recent collider experiments is the finding that hadronic total cross sections as well as elastic cross sections can be described by the exchange of a "soft Pomeron" pole in the near-forward limit. The most important consequence of Pomeron being a pole is the factorization property. However, due to Pomeron intercept being greater than 1, the extrapolated single diffraction dissociation cross section based on a classical triple-Pomeron formula is too large leading to a potential unitarity violation at Tevatron energies. It is our desire here to examine carefully the notion of Pomeron flux renormalization, proposed by Goulianos, and to suggest an alternative resolution. Our treatment lies in a proper implementation of final-state screening correction, with "flavoring" for Pomeron as the primary dynamical mechanism for setting the relevant energy scale. In our scheme, which we shall occasionally refer to as "unitarization of Pomeron flux", initial-state screening remains unimportant, consistent with the pole dominance picture for elastic and total cross section hypothesis at Tevatron energies. Since factorization remains intact, umambiguous predictions for double Pomeron exchange, doubly diffraction dissociation, etc., both at Tevatron and at LHC energies, can be made.

1 Introduction

One of the more interesting developments from recent collider experiments is the finding that hadronic total cross sections as well as elastic cross sections in the near-forward limit can be described by the exchange of a "soft Pomeron" pole, [1] *i.e.*, the absorptive part of the elastic amplitudes can be approximated by $Im\, T_{a,b}(s,t) \simeq \beta_a(t) s^{\alpha_{\mathcal{P}}(t)} \beta_b(t)$. The Pomeron trajectory has two important features. [2] First, its zero-energy intercept is greater than one, $\alpha_{\mathcal{P}}(0) \equiv 1+\epsilon$, $\epsilon \simeq 0.08 \sim 0.12$, leading to rising $\sigma^{tot}(s)$. Second, its Regge slope is approximately $\alpha'_{\mathcal{P}} \simeq 0.25 \sim 0.3\ GeV^{-2}$, leading to the observed shrinkage effect for elastic peaks. The most important consequence of Pomeron being a pole is factorization. For a singly diffractive dissociation process, factorization leads to a "classical triple-Pomeron" formula, [4]

$$\frac{d\sigma}{dtd\xi} \to \frac{d\sigma^{classical}}{dtd\xi} \equiv F^{cl}_{\mathcal{P}/a}(\xi,t)\sigma^{cl}_{\mathcal{P}b}(M^2,t), \tag{1}$$

where M^2 is the missing mass variable and $\xi \equiv M^2/s$. The first term on the right-hand side of Eq. (1) is the so-called "Pomeron flux", and the second term is the "Pomeron-particle" total cross section. Eq. (1) is in principle valid only when ξ^{-1} and M^2 are both large, with t small and held fixed. However, with $\epsilon \sim 0.1$, it has been observed [5] that the extrapolated $p\bar{p}$ single diffraction dissociation cross section, σ^{sd}, based on the standard triple-Pomeron formula is too large at Tevatron energies by as much as a factor of $5 \sim 10$ and it could become larger than the total cross section.

Let us denote the singly diffractive cross section as a product of a "renormalization" factor and the classical formula,

$$\frac{d\sigma}{dtd\xi} = Z(\xi, t; s)\frac{d\sigma^{classical}}{dtd\xi}. \tag{2}$$

It was argued by K. Goulianos in Ref. [5] that agreement with data could be achieved by having an energy-dependent suppression factor, $Z(\xi, t; s) \to Z_G(s) \equiv N(s)^{-1} \leq 1$. This "renormalization" factor is chosen so that the new "Pomeron flux", $F_N(s, \xi, t) \equiv N(s)^{-1}F^{cl}_{\mathcal{P}/p}(\xi, t)$, is normalized to unity for $s \geq \bar{s}$, $\sqrt{\bar{s}} \simeq 22\, GeV$.[5,6,7] However, the modified triple-Pomeron formula no longer has a factorized form. An alternative suggestion has been made recently by P. Schlein.[8,9] It was argued that phenomenologically, after incorporating lower triple-Regge contributions, the renomalization factor for the triple-Pomeron contribution could be described by an ξ-dependent suppression factor, $Z \to Z_S(\xi)$.

In view of the factorization property for total and elastic cross sections, the "flux renormalization" procedure appears paradoxical and could undermine the theoretical foundation of a soft Pomeron as a Regge pole from a non-perturbative QCD perspective. We shall refer to this as "**Dino's paradox**". Finding a resolution that is consistent with Pomeron pole dominance for elastic and total cross sections at Tevatron energies will be the main focus of this study. In particular, we want to maintain the following factorization property,

$$\frac{d\sigma}{dtd\xi} \to \sum_k D_k(\xi, t)\Sigma_k(M^2), \tag{3}$$

when ξ^{-1} and M^2 become large.[11] It is our intention to provide a solution without deviating from generally accepted guiding principles for hadron dynamics.

It is reasonable to expect that the resolution to the paradox should lie in a proper implementation of screening corrections to the classical triple-Pomeron formula. One of our key observations is the fact that, since σ^{sd} is approximately 15% of the total cross section at Tevatron, a pole dominance picture for σ^{tot} should also impose a direct constraint on the nature of diffractive production. The second observation is the necessity in having an overall coherent scheme in which various different hadronic energy scales must enter consistently. In our usage of classical triple-Regge formulas, the basic energy scale is always in terms of $s_0 \simeq 1$

GeV^2. We demonstrate that there are at least two other energy scales, $s_r \equiv e^{y_r}$ and $s_f \equiv e^{y_f}$, $y_r \simeq 3 \sim 5$ and $y_f \simeq 8 \sim 10$, which must be incorporated properly. The first is associated with the physics of light quarks and the scale of chiral symmetry breaking. The second is the "flavoring" scale and is associated with "heavy flavor" production. In a non-perturbative QCD setting, both play an important role in our understanding of a bare Pomeron with an intercept greater than unity. [12,13]

Our treatment lies in a proper implementation of final-state screening correction, (or **final-state unitarization**), with "**flavoring**" for Pomeron as the primary dynamical mechanism for setting the relevant energy scale. In our treatment, initial-state screening remains unimportant, consistent with the pole dominance picture for elastic and total cross section hypothesis at Tevatron energies. The factor $D_{a,\mathcal{P}}(\xi,t)$ from the Pomeron contribution will be referred to as a "unitarized Pomeron flux factor," and we shall occassionally refer to our procedure as "**unitarization of Pomeron flux**". We shall demonstrate that in our unitarization scheme the total renormalization factor has a factorized form,

$$Z(\xi,t;s) = Z_d(\xi,t)Z_m(M^2) = [S_f(\xi,t)R^2(\xi^{-1})]R(M^2), \tag{4}$$

where S_f is due to final-state screening, Eq. (38). R is a flavoring factor, given by Eq. (31), and there is one flavoring factor for each Pomeron propagator.

With the pole dominance hypothesis, we demonstrate that the unitarized flux factor, $D_{a,\mathcal{P}}(\xi,t) \equiv Z_d(\xi,t)F^{cl}_{a,\mathcal{P}}(\xi,t)$, must satisfy a normalization condition.

$$\int_{-\infty}^{0} dt \int_0^1 d\xi D_{a,\mathcal{P}}(\xi,t)\xi^{\epsilon} \equiv \beta_a^{diff} < \beta_a(0). \tag{5}$$

The required damping to overcome the $\xi^{-(1+2\epsilon)}$ divergent behavior of $F^{cl}_{a,\mathcal{P}}(\xi,0)$ at small ξ comes from both the screening factor $S_f(\xi,t)$ and the factor ξ^{ϵ} from the "Pomeron-particle" total cross section. More importantly, we point out that, even without screening, flavoring alone would lead to a suppression for the large ξ^{-1} and large M^2 region at Tevatron energies of the order $\{R(\infty)/R(1)\}^3 \simeq e^{-3\epsilon y_f} \simeq 1/8$, where $R(\infty) = 1$. For instance, flavoring leads to a suppression for the triple-Pomeron coupling

$$g_{\mathcal{PPP}}(0) \sim e^{-3\epsilon y_f/2} g^{low}_{\mathcal{PPP}}(0), \tag{6}$$

where $g^{low}_{\mathcal{PPP}}(0) \simeq 0.364 \pm 0.025\ mb^{\frac{1}{2}}$ is that obtained in a fit to low-energy data. [14] In other words, the apparent unitarity violation will be mostly alleviated by flavoring alone! Lastly we emphasize, since factorization remains intact, our unitarization scheme leads to unambiguous predictions in terms of these suppression factors for double Pomeron exchange, doubly diffraction dissociation, etc., both at Tevatron and at LHC energies.

In Section II, we explain why, given the Pomeron pole dominance hypothesis, the initial-state screening cannot be large at Tevatron energies. We introduce the notion of "rapidity gap" cross sections and explain the sum rule, Eq. (5). We study in Sec. III the effect of final-state absorption. In particular, we demonstrate that, with Pomeron intercept greater than one, absorption is "total" within the "expanding hadronic disk" and the residual inelastic scattering can only take place on the "edge" of the disk. It follows that unitarity will not be violated by diffraction dissociation within a pole dominance picture for σ^{tot}. A model for calculating the screening factor, $S_f(\xi, t)$, is introduced. In Sec. IV we discuss the origin of the flavoring energy scale and indicate its relevant for diffractive dissocition cross sections. In a scale-dependent formulation, we explain how flavoring contributes factors $R^2(\xi^{-1})$ and $R(M^2)$ to $Z(\xi, t; s)$ in order to account for both the Pomeron pole and coupling renormalizations. The flavoring function R can be expressed in terms of a scale-dependent "effective" Pomeron intercept, $\epsilon_{eff}(y)$.

Putting these together, we provide the **final recipe** for our resolution to Dino's paradox in Sec. V. Phenomenologically simple parametrizations for both the screening function, S_f, and the effective intercept, ϵ_{eff}, are presented. Readers interested only in our proposed modification to the classical triple-Pomeron formula can go there directly. We present a phenomenological analysis which yields an estimate for the "high energy" triple-Pomeron coupling:

$$g_{\mathcal{PPP}}(0) \simeq .14 \sim .20 \ mb^{\frac{1}{2}}. \tag{7}$$

This value is consistent with our flavoring expectation, Eq. (6). Surprisingly, the amount of screening required at Tevatron energies seems to be very small. Predictions fro other related cross sections are discussed in Sec. VI. Comparison of our approach to that of Refs. [5] and [8] as well as other comments are given in Sec. VII.

2 Pomeron Dominance Hypothesis at Tevatron Energies

We shall first explore consequences of the observation that both total cross sections and elastic cross sections can be well described by a Pomeron pole exchange at Tevatron energies. Absorption correction, if required, seems to remain small. Since the singly diffractive cross section, σ^{sd}, is a sizable part of the total, it must also grow as s^{ϵ}. This qualitative understanding can be quantified in terms of a sum rule for "rapidity gap" cross sections. This in turn imposes a convergence condition on our unitarized Pomeron flux, $D_{a,\mathcal{P}}(\xi, t)$.

To simplify the discussion, we shall first ignore transverse momentum distribution by treating the longitudinal phase space only. For instance, for singly diffraction dissociation, the longitudinal phase space can be specified by two rapidities, $y \equiv \log(\xi^{-1})$ and $y_m \equiv \log M^2$. The first variable specifies the rapidity gap associated with the detected leading proton (or antiproton), and the second variable specifies the rapidity "span" of the missing mass distribution. At fixed s, they are constrained by $y + y_m \simeq Y \equiv \log s$, (see Figure 1), and we can speak of differential diffractive cross section $d\sigma^{sd}/dy$. We shall in what follows use $\{\xi^{-1}, M^2\}$ and $\{y \equiv \log \xi^{-1}, y_m \equiv \log M^2\}$ interchangably. Dependence on transverse degrees of freedom can be re-introduced without much effort after completing the main discussion.

Consider the process $a + b \to c + X^<$, where the number of particles in X is unspecified. However, unlike the usual single-particle inclusive process, the superscript for $X^<$ indicates that all particles in X must have rapidity less than that of the particle c, *i.e.*, the detected particle c is the one in the final state with the largest rapidity value. Kinematically, a single-gap cross section is identical to the singly diffraction dissociation cross section discussed earlier. Under the assumption where all transverse motions are unimportant, one has $y_c \simeq y_a \equiv Y_{max}$ and the differential gap cross section, $d\sigma^{gap}/dy$, can also be considered as a function of y and y_m, with $y + y_m \simeq Y$.

Because the detected particle c has been singled out to be different from all other particles, this is no longer an inclusive cross section, and it does not satisfy the usual inclusive sum rules. Upon integrating over the rapidity gap y and summing over particle type c, no multiplicity enhancement factor is introduced and one obtains simply the total cross section, *i.e.*, a gap cross section satisfies the following exact sum rule,

$$\sum_c \int dy \frac{d\sigma^{gap}_{ab\to c}}{dy} \equiv \sum_c \sigma^{gap}_{ab\to c} = \sum_{n=2}^{\infty} \sigma_n = \sigma^{tot}_{ab}. \tag{8}$$

Interestingly, this allows an identification of the total cross section as a sum over specific gap cross sections, $\sigma^{gap}_{ab\to c}$, each is "derived" from a specific "leading particle" gap distribution. Since no restriction has been imposed on the nature of the "gap distribution", *e.g.*, particle c can have different quantum numbers from a, the notion of a gap cross section is more general than a diffraction cross section.[15] It follows that the singly diffractive dissociation cross section, σ^{sd}_{ab}, is a part of $\sigma^{gap}_{ab\to a}$.

Consider next our factorized ansatz, Eq. (3). For y and y_m large, it leads to a gap distribution, $d\sigma^{gap}_{ab\to c}/dy = e^{-y} D_{a\to c}(y)\Sigma_b(y_m)$. If $\Sigma_b(y_m) = \beta_b e^{\epsilon y_m}$, it follows that contribution from each gap distribution is Regge behaved, $\sigma^{gap}_{ab\to c} \simeq \beta^c_a c^{\epsilon Y} \beta_b$, where the total Pomeron residue is a sum of "partial residues",

$$\beta_a = \sum_c \beta^c_a = \sum_c \int_0^\infty dy D^c_a(y) e^{-(1+\epsilon)y}, \tag{9}$$

For above integral to converge, each flux factor must grow slower than $e^{\epsilon y}$. That is, $D^c_a(y)e^{-\epsilon y} \to 0$ as $y \to \infty$. In a traditional Regge approach, the large rapidity gap behavior for each flux factor is controlled by an approriate Regge propagator, $e^{(\alpha_i+\alpha_j-1)y}$. Clearly a standard triple-Pomeron behavior with $\alpha_{\mathcal{P}} > 1$ is inconsistent with the pole dominance hypothesis. Unitarity correction must supply enough damping to provide convergence.

There is yet another way of expressing the consequence of the pole dominance hypothesis. Dividing each gap differential cross section by the total cross section, factorization of Pomeron leads to a "limiting distribution": $\rho_{ab}(y, Y) \to \rho_a(y) \equiv \sum_c \rho_a^c(y)$. That is, the limit is independent of the total rapidity, Y, and the gap density is normalizable, $\int_0^\infty dy \rho_a(y) = \sum_c \int_0^\infty dy \rho_a^c(y) = \sum_c (\beta_a^c/\beta_a) = 1$. This normalization condition for the gap distribution is precisely Eq. (9).

Let us now restore the transverse distribution and concentrate on the diffractive dissociation contribution, which can be identified with the high M^2 and high ξ^{-1} limit of $d\sigma_{ab\to a}^{gap}(t, \xi; M^2)$. In terms of ξ, M^2, and t, the differential cross section at large M^2 under our factorizable ansatz takes on the following form, $d\sigma_{ab}^{sd}/dtd\xi \simeq D_{a,\mathcal{P}}(\xi, t)\Sigma_{\mathcal{P},b}^{cl}(M^2)$, where $\Sigma_{\mathcal{P},b}^{cl}(M^2) = (M^2)^\epsilon \beta_b(0)$. It follows from Eq. (9) that $D_{a,\mathcal{P}}(\xi, t)$ must satisfy the following bound:

$$\int_{-\infty}^{0} dt \int_{\xi_{min}(s)}^{\xi_{max}} d\xi D_{a,\mathcal{P}}(\xi, t)\xi^\epsilon \leq \int_{-\infty}^{0} dt \int_0^1 d\xi D_{a,\mathcal{P}}(\xi, t)\xi^\epsilon \equiv \beta_a^{diff} < \beta_a(0). \tag{10}$$

The hypothesis of a Pomeron pole dominance for the total and elastic cross sections is of course only approximate. However, to the extend that absorptive corrections remain small at Tevatron energies, one finds that a modified Pomeron flux factor must differ from the "classical" Pomeron flux at small ξ in such a way so that the upper bound in Eq. (10) is satisfied. We shall refer to $D_{a,\mathcal{P}}(\xi, t)$ as the "unitarized Pomeron flux". How this can be accomplished via final state screening will be discussed next. Note both the **similarity** and the **difference** between Eq. (10) and the "flux normalization" condition mentioned in the Introduction. Here, this convergent integral yields a finite number, β_a^{diff}, and the ratio $\beta_a^{diff}/\beta_a(0)$ can be interpreted as the probability of having a diffractive gap at high energies.

3 Final-State Screening

The best known example for implementing the idea of "screening" in high energy hadronic collisions has been the "expanding disk" picture for rising total cross sections. Diffraction scattering as the shadow of inelastic production has been a well established mechanism for the occurence of a forward peak. Analyses of data upto collider energies have revealed that the essential feature of nondiffractive particle production can be understood in terms of a multipertipheral cluster-production mechanism. In such a picture, the forward amplitude is predominantly absorptive and is dominated by the exchange of a "bare Pomeron". If the Pomeron intercept is greater than one, it forces further unitarity corrections as one moves to higher energies. For instance, saturation of the Froiossart bound can be next understood through an eikonel mechanism, with the absorptive eikonal $\chi(s, b)$ given by the bare Pomeron amplitude in the impact-parameter space.

The main problem we are facing here is not so much on how to obtain a "most accuraate" flux factor $D_{a,\mathcal{P}}(\xi,t)$ at very small ξ. We are concerned with a more difficult conceptual problem of how to reconcile having a potentially large screening effect for diffraction dissociation processes and yet being able to maintain approximate pole dominance for elastic and total cross sections up to Tevatron energies. We shall show using an expanding disk picture that absorption works in such a way that inelastic scattering can only take place on the "edge" of disk.[16] Therefore, once applied using final-state screening, the effect of initial-state absorption will be small, hence allowing us to maintain Pomeron pole factorization for elastic and total cross sections.

3.1 Expanding Disk Picture

Let us briefly review this picture which also serves to establish notations. At high energies, a near-forward amplitude can be expressed in an impact-parameter representaion via a two-dimensional Fourier transform,

$$T(s,t) \equiv 2is \int d^2\vec{b} e^{i\vec{q}\cdot\vec{b}} \tilde{f}(s,b), \qquad \tilde{f}(s,b) = (4i\pi s)^{-1} \int d^2\vec{q} e^{-i\vec{b}\cdot\vec{q}} T(s,t), \tag{11}$$

where $t \simeq -\vec{q}^2$. Assume that the near-forward elastic amplitude at moderate energies can be described by a Born term, *e.g.*, that given by a single Pomeron exchange where we shall approximate it to be purely absorptive. Let us denote the contribution from the Pomeron exchange to $\tilde{f}(s,b)$ as $\chi(s,b)$. With $\alpha_{\mathcal{P}}(t) = 1 + \epsilon + \alpha'_{\mathcal{P}} t$, and approximating $\beta(t)$ by an exponential, we find $\chi(s,b) \simeq X(s) e^{-(b^2/4B(s))}$,

$$X(s) \equiv \chi(s,0) \simeq \frac{\sigma_0(s)}{4B(s)} \tag{12}$$

where $B(s) = b_0 + \alpha'_{\mathcal{P}} \log s$ and $\sigma_0(s) = \sigma_0 s^{\epsilon}$. With $\epsilon > 0$, the Born approximation would eventually violate s-channel unitarity at small b as s increases. A systematic procedure which in principle should restore unitarity is the Reggeon calculus. However, our current understanding of dispersion-unitarity-relation is too qualitative to provide a definitive calculational scheme.

The key ingredient of "screening" correction is the recognition that the next order correction to the Born term must have a negative sign. (The sign of double-Pomeron cut contribution.) In an impact-representation, Reggeon calculus assures us that the correction can be represented as

$$\tilde{f}_1(s,b) \simeq -\frac{1}{2!}\, \mu(s)\, \chi(s,b)^2, \tag{13}$$

where μ is positive. To go beyond this, one needs a model. A physically well-motivated model which should be meaningful at moderate energies and allows easy analytic treatment is the eikonal model. Writing $\tilde{f}(s,b) = \tilde{f}_0(s,b) + \tilde{f}_1(s,b) + \tilde{f}_2(s,b) + \cdots$, the expansion alternates in sign, and with simple weights such that $\tilde{f}(s,b) = [1 - e^{-\mu\chi(s,b)}]/\mu$, and

$$T(s,t) = \left(\frac{2is}{\mu}\right) \int d^2\vec{b}\, e^{i\vec{q}\cdot\vec{b}} \{1 - e^{-\mu\chi(s,b)}\}. \tag{14}$$

Conventional eikonal model has $\mu = 1$. We keep $\mu \leq 1$ here so as to allow the possibility that screening is "imperfect".

Observe that the eikonal derived from the Pomeron exchange, $\chi(s,b)$, is a monotonically decreasing function of b^2, taking on its maximum value $X(s)$ at $b^2 = 0$, which increases with s due to $\epsilon > 0$. The eikonal drops to zero at large b^2 and is of the order 1 at a radius, $b_c(s) \simeq \sqrt{B(s) \log \mu X(s)} \sim \log s$. Within this radius, $\tilde{f}(s,b) = O(1)$ and it vanishes beyond. This is the "expansion disk" picture of high energy scattering, leading to an asymptotic total cross section $O(b_c(s)^2)$.

3.2 Inelastic Screening

In order to discuss inelastic final-state screening, we follow the "shadow" scattering picture in which the "minimum biased" events are predominantly "short-range ordered" in rapidity and the production amplitudes can be described by a multiperipheral cluster model. Substituting these into the right-hand side of an elastic unitary equation, $ImT(s,0) = \sum_n |T_{2,n}|^2$, one finds that the resulting elastic amplitude is dominated by the exchange of a Regge pole, which we shall provisionally refer to as the "bare Pomeron". Next consider singly diffractive events. We assume that the "missing mass" component corresponds to no gap events, thus the distribution is again represented by a "bare Pomeron". However, for the gap distribution, one would insert the "bare Pomeron" just generated into a production amplitude, thus leading to the classical triple-Pomeron formula.

Extension of this scheme leads to a "perturbative" treatment for the total cross section in the number of bare Pomeron exchanges along a multiperipheral chain. Such a scheme was proposed long time ago, [18] with the understanding that the picture could make sense at moderate energies, provided that the the intercept of the Pomeron is one, $\alpha(0) \simeq 1$, or less. However, with the acceptance of a Pomeron having an intercept greater than unity, this expansion must be embellished. Although it is still meaningful to have a gap expansion, one must improve the descriptions for parts of a production amplitude involving large rapidity gaps by taking into account absorptions for the gap distribution. We propose that this "partial unitarization" be done for each gap separately, thus maintaining the factorization property along each short-range ordered sector. This involves final-state screening, and, for singly diffraction dissociation, it corresponds to the inclusion of "enhanced Pomeron diagrams" in the triple-Regge region.

To simplify the notation, we shall use the energy variable, ξ^{-1}, and the rapidity gap variable, $y = \log(\xi^{-1})$, interchangeably. Let's express the total unitarized contribution to a gap cross section in terms of a "unitarized flux" factor, $D(y,t) \equiv (e^y/16\pi)|g_d(y,t)|^2 \equiv (1/16\pi\xi)|f_d(\xi,t)|^2 g_{\mathcal{PPP}}(t)$, so that it reduces to the classical triple-Pomeron formula as its Born term. That is, the corresponding Born amplitude for $f_d(\xi,t)$ is the "square-root" of the triple-Pomeron contribution to the classical formula, $f_{\mathcal{P}}^{cl}(\xi,t) = \beta_{\mathcal{P}}(t)(\xi^{-1})^{(\alpha_{\mathcal{P}}(t)-1)}$. Screening becomes important if large gap becomes favored, *i.e.*, when $\epsilon > 0$.

Let us work in an impact representation, $g(y_d,t) \equiv 2i\int d^2\vec{q}e^{i\vec{b}\cdot\vec{q}}\tilde{g}_d(y_d,b)$. Consider an expansion $\tilde{g}(y,b) = \chi_d(y,b) + \tilde{g}_1(y,b) + \tilde{g}_2(y,b) + \cdots$ where, under the usual exponential approximation for the t-dependence, the Fourier transform for the Born term is $\chi_d(y,b) = \frac{\sigma^d(y)}{4B_d(y)}e^{-\frac{b^2}{4B_d(y)}}$, with $B_d(y) = b_d + \alpha'_{\mathcal{P}}y$ and $\sigma^d(y) = \sigma_d e^{\epsilon\, y}$. The key physics of absorption is

$$\tilde{g}_1(y,b) \simeq -\mu_d\chi(y,b)\chi_d(y,b). \tag{15}$$

The proportionality constant μ_d can be different from the constant μ introduced earlier for the elastic screening, either for kinematic or dynamic reasons, or both. For a generalized eikonal approximation, one has $\tilde{g}(y,b) = \chi_d\{1 - \mu_d\chi + \frac{1}{2!}(\mu_d\chi)^2 - \cdots\} = \chi_d(y,b)e^{-\mu_d\,\chi(y,b)}$. If we define the "final-state screening" factor as the ratio between the unitarized flux factor and the classical triple-Pomeron formula, $D_{a,\mathcal{P}}(y,t) = S_f^{(0)}(y,t;X)F_{a,\mathcal{P}}^{cl}(y,t)$, we then have

$$S_f^{(0)}(y,t;X) = |f_d(y,t)/f_d^{cl}(y,t)|^2. \tag{16}$$

We shall use this expression as a model for probing the physics of inelastic screening in an expanding disk picture.

Let us examine this eikonal screening factor in the forward limit, $t = 0$, where

$$S_f^{(0)}(y,0;X)^{\frac{1}{2}} = \frac{\int db^2\chi_d(y,b)e^{-\mu_d\chi(y,b)}}{\int db^2\chi_d(y_d,b)}. \tag{17}$$

Unlike the elastic situation, the integrand of the numerator is strongly suppressed both in the region of large b^2 and in the region "inside" the expanding disk. **The only significant contribution comes from a "ring" region near the edge of the expanding disk.** [16,19] Since the value of the integrand is of $O(1)$ there, one finds that the numerator varies with energy only weakly. On the other hand, the denominator is simply $\sigma^d(y)$, which increases as $e^{\epsilon y}$. Therefore, this leads to an exponential cutoff in y, $S_f^{(0)}(y,0;X) \sim e^{-2\epsilon y}$. This damping factor precisely cancels the $\xi^{-2\epsilon}$ behavior from the classical triple-Pomeron formula at small ξ, leading to a unitarized Pomeron flux factor.

To be more precise, let us work in a simpler representation for $S_f^{(0)}(y,0;X)$ by changing the variable $b^2 \to z \equiv e^{-b^2/4B(y)}$. With our gaussian approximation, one finds that

$$S_f^{(0)}(y,0;X) = \{r\int_0^1 dz z^{r-1}e^{-xz}\}^2|_{x=\mu_d X(y),r=r(y)}, \tag{18}$$

where $r(y) \equiv B(y)/B_d(y)$. This expression can be expressed as $\{rx^{-r}\Gamma(x,r)\}^2|_{x=\mu_d X(y), r=r(y)}$, where $\Gamma(x,r) = \int_0^x dz z^{r-1} e^{-z}$ is the incomplete Gamma function. In this representation, one easily verifies that the screening factor has the desired properties: As $\mu_d \to 0$, screening is minimal and one has $S_f^{(0)}(y, 0; X) \to 1$. On the other hand, for y large, $X(y)$ increases so that $S_f^{(0)}(y, 0; X) \to [\mu_d X(y)]^{-2}$, as anticipated.

Similarly, we find that the logrithmic width for the unitarized flux $D(y,t)$ at $t = 0$ has increased from $2B_d$ to $2B_d^{eikonal}$ where $B_d^{eikonal} = B_d\{-r\frac{d}{dr}\log\int_0^1 dz z^{r-1} e^{-xz}\}_{x=\mu_d X(y), r=r(y)}$. As $\mu_d \to 0$, $B_d^{eikonal} \to B_d$. For y very large, $B_d^{eikonal} \to B \log \mu_d X(y) \sim b_c(y)^2 \propto y^2$. This corresponds to a faster shrinkage than that of ordinary Regge behavior. Averaging over t, one finds at large diffractive rapidity y, the final-state screening provides an avearge damping

$$\langle S_f \rangle \to e^{-2\epsilon\, y} = \xi^{2\epsilon}. \tag{19}$$

This leads to a unitarized Pomeron flux, $D_{a,\mathcal{P}}(\xi, t)$, which automatically satisfies the upper bound, Eq. (10), derived from the Pomeron pole dominance hypothesis.

4 Dynamics for Soft Pomeron and Flavoring

Although final-state screening would automatically avoid unitarity violation, the primary source of high energy suppression actually comes from a proper treatment of **scale-dependence** for Pomeron couplings. Consider for the moment the following scenario where one has two different fits to hadronic total cross sections:

- (a) "High energy fit": $\sigma_{ab}(Y) \simeq \beta_a\, \beta_b\, e^{\epsilon\, Y}$ for $Y >> y_f$,
- (b) "Low energy fit": $\sigma_{ab}(Y) \simeq \beta_a^{low}\, \beta_b^{low}$ for $0 < Y << y_f$.

That is, we envisage a situation where the "effective Pomeron intercept", ϵ_{eff}, changes as one moves up in energies. Assuming a smooth interpolation between these two fits, one can obtain the following order of magnitude estimate

$$\beta_p \simeq e^{-\frac{\epsilon\, y_f}{2}} \beta_p^{low}. \tag{20}$$

Modern parametrization for Pomeron residues typically leads to values of the order $(\beta_p)^2 \simeq 14 \sim 17$ mb. However, before the advent of the notion of a Pomeron with an intercept greater than 1, a typical parametrization would have a value $(\beta_p^{low})^2 \simeq 35 \sim 40$ mb, accounting for a near constant Pomeron contribution at low energies. This leads to an estimate of $y_f \sim 8$, corresponding to $\sqrt{s} \sim 50$ GeV. This is precisely the energy scale where a rising total cross section first becomes noticeable.

The scenario just described has been referred to as "flavoring", the notion that the underlying effective degrees of freedom for Pomeron will increase as one moves to higher energies,[21,22,23] and it has provided a dynamical basis for understanding the value of Pomeron intercept in a non-perturbative QCD setting.[12,13] In this scheme, both the Pomeron intercept and the Pomeron residues are **scale-dependent**. We shall briefly review this mechanism and introduce a scale-dependent formalism where the entire flavoring effect can be absorbed into a flavoring factor, $R(y)$, associated with each Pomeron propagator.

4.1 Bare Pomeron in Non-Perturbative QCD

In a non-perturbative QCD setting, the Pomeron intercept is directly related to the strength of the short-range order component of inelastic production and this can best be understood in a large-N expansion.[24,25] In such a scheme, particle production mostly involves emitting "massless pions", and the basic energy scale of interactions is that of ordinary vector mesons, of the order of 1 GeV. In a one-dimensional multiperipheral realization for the "planar component" of the large-N QCD expansion, the high energy behavior of a n-particle total cross section is primarily controlled by its longitudinal phase space, $\sigma_n \simeq (g^4 N^2/(n-2)!)(g^2 N \log s)^{n-2} s^{J_{eff}-1}$. Since there are only Reggeons at the planar diagram level, one has $J_{eff} = 2\alpha_R - 1$ and, after summing over n, one arrives at Regge behavior for the planar component of σ^{tot} where

$$\alpha_R = (2\alpha_R - 1) + g^2 N. \tag{21}$$

At next level of cylinder topology, the contribution to partial cross section increases due to its topological twists, $\sigma_n \simeq (g^4/(n-2)!)2^{n-2}(g^2 N \log s)^{n-2} s^{J_{eff}-1}$, and, upon summing over n, one arrives at a total cross section governed by a Pomeron exhange,

$$\sigma_0^{tot}(Y) = g^4 e^{\alpha_{\mathcal{P}} Y} \tag{22}$$

where the Pomeron interecept is

$$\alpha_{\mathcal{P}} = (2\alpha_R - 1) + 2g^2 N. \tag{23}$$

Combining Eq. (21) and Eq. (23), we arrive at the following amazing "bootstrap" result,

$$\alpha_{\mathcal{P}} \simeq 1. \tag{24}$$

In a non-perturbative QCD setting, having a Pomeron intercept near 1 therefore depends crucially on the topological structure of large-N non-Abelian gauge theories.[12,13,24,25] In this picture, one has $\alpha_R \simeq .5 \sim .7$ and $g^2N \simeq .3 \sim .5$. With $\alpha' \simeq 1\ GeV^{-2}$, one can also directly relate α_R to the average mass of typical vector mesons. Since vector meson masses are controlled by constituent mass for light quarks, and since constituent quark mass is a consequence of chiral symmetry breaking, the Pomeron and the Reggeon intercepts are directly related to fundamental issues in non-perturbative QCD.

Finally we note that, in a Regge expansion, the relative importance of secondary trajectories to the Pomeron is controlled by the ratio $e^{\alpha_R\, y}/e^{\alpha_{\mathcal{P}}\, y} = e^{-(\alpha_{\mathcal{P}}-\alpha_{\mathcal{R}})\, y}$. It follows that the relevant scale in rapidity we are seeking is precisely $(\alpha_{\mathcal{P}} - \alpha_R)^{-1} < y_r \simeq 3 \sim 5$. The importance of this scale y_r is of course well known: When using a Regge expansion for total and two-boby cross sections, secondary trajactory contributions become important and must be included whenever rapidity separations are below $3 \sim 5$ units. This is of course also true for the triple-Regge region. That is, in terms of y and y_m, triple-Regge terms other than the leading triple-Pomeron term must be added whenever one moves into regions when $y \leq y_r$ or $y_m \leq y_r$ or both.[8,9]

4.2 Flavoring of Bare Pomeron

We have argued previously that "heavy flavor" production provides an additional energy scale, $s_f = e^{y_f}$, for soft Pomeron dynamics, and the effect of heavy flavors can be responsible for the perturbative increase of the Pomeron intercept to be greater than unity, $\alpha_{\mathcal{P}}(0) \sim 1+\epsilon$, $\epsilon > 0$. One must bear this additional energy scale in mind in working with a soft Pomeron.[12,13,21,22,23] That is, to fully justify using a Pomeron with an intercept $\alpha_{\mathcal{P}}(0) > 1$, one must restrict oneself to energies $s > s_f$ where heavy flavor production is no longer suppressed. Conversely, to extrapolate Pomeron exchange to low energies below s_f, a lowered "effective trajectory" must be used. This feature of course is unimportant for total and elastic cross sections at Tevatron energies. However, it is important for diffractive production since both ξ^{-1} and M^2 will sweep right through this energy scale at Tevatron energies. (See Figure 1.)

Flavoring becomes important whenever there is a further inclusion of effective degrees of freedom than that associated with light quarks. This can again be illustrated by a simple one-dimensional multiperipheral model. In addition to what is already contained in the Lee-Veneziano model, suppose that new particles can also be produced in a multiperipheral chain. Concentrating on the cylinder level, the partial cross sections will be labelled by two indices,

$$\sigma_{p,q} \simeq (g^4/p!q!)2^{p+q}(g^2N\log s)^p(g_f^2N\log s)^q s^{J_{eff}-1}, \tag{25}$$

where q denotes the number of clusters of new particles produced. Upon summing over p and q, we obtain a "renormalized" Pomeron trajectory

$$\alpha_{\mathcal{P}} = \alpha_{\mathcal{P}}^{old} + \epsilon, \tag{26}$$

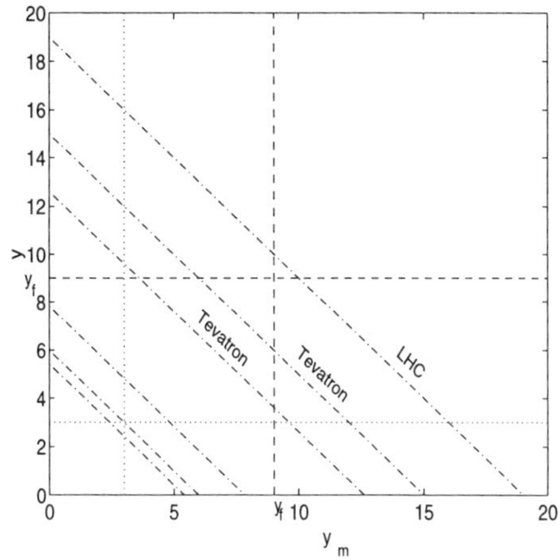

Figure 1: Phase space for single diffraction dissociation from ISR to LHC in terms of rapidity variables $y \equiv \log \xi^{-1}$ and $y_m \equiv \log M^2$. The dashed lines are for "flavoring" scale, y_f, chosen to be 9 for illustration. Dotted lines are $y_{min} \simeq 3$ lines for both y and y_m. The Dashed-dotted lines are constant center of mass energy lines for E_i, $i = 1, \cdots\cdots, 6$, equals to $15, 30, 60, 630, 1800, 14,000$ GeV respectively.

where $\alpha_{\mathcal{P}}^{old} \simeq 1$ and $\epsilon \simeq 2g_f^2 N$. That is, in a non-perturbative QCD setting, the intercept of Pomeron is a dynamical quantity, reflecting the effective degrees of freedom involved in near-forward particle production.[12,13,26]

If the new degree of freedom involves particle production with high mass, the longitudinal phase space factor, instead of $(\log s)^q$, must be modified. Consider the situation of producing one $N\bar{N}$ bound state together with pions, *i.e.*, p arbitrary and $q = 1$ in Eq. (25). Instead of $(\log s)^{p+1}$, each factor should be replaced by $(\log(s/m_{eff}^2))^{p+1}$, where m_{eff} is an effective mass for the $N\bar{N}$ cluster. In terms of rapidity, the longitudinal phase space factor becomes $(Y - \delta)^{p+1}$, where δ can be thought of as a one-dimensional "excluded volume" effect. For heavy particle production, there will be an energy range over which $q = 1$ remains a valid approximation. Upon summing over p, one finds that the additional contribution to the total cross section due to the production of one heavy-particle cluster is[21,22]

$$\sigma_{q=1}^{tot} \sim \sigma_0^{total}(Y - \delta)(2g_f^2 N)\log(Y - \delta)\theta(Y - \delta), \tag{27}$$

where $\alpha_{\mathcal{P}}^{old} \simeq 1$.

Note the effective longitudinal phase space "threshold factor", $\theta(Y - \delta)$, and, initially, this term represents a small perturbation to the total cross section obtained previously, (corresponding to $q = 0$ in Eq. (25)), σ_0^{total}. Over a rapidity range, $[\delta, \delta+\delta_f]$, where δ_f is the average rapidity required for producing another heavy-mass cluster, this is the only term needed for incorporating this new degree of freedom. As one moves to higher energies, "longitudinal phase space suppression" becomes less important and more and more heavy particle clusters will be produced. Upon summing over q, we would obtain a new total cross section, described by a renormalized Pomeron, with a new intercept given by Eq. (26).

We assume that, at Tevatron, the energy is high enough so that this kind of "threshold" effects is no longer important. How low an energy do we have to go before one encounter these effects? Let us try to answer this question by starting out from low energies. As we have stated earlier, for $Y > 3 \sim 5$, secondary trajectories become unimportant and using a Pomeron with $\alpha \simeq 1$ becomes a useful approximation. However, as new flavor production becomes effective, the Pomeron trajectory will have to be renormalized. We can estimate for the relevant rapidity range when this becomes important as follows: $y_f > 2\delta_0 + < q >_{min} \delta_f$. The first factor δ_0 is associated with leading particle effect, i.e., for proton, this is primariy due to pion exchange. δ_f is the minimum gap associated with one heavy-mass cluster production, *e.g.*, nucleon-antinuceon pair production. We estimate $\delta_0 \simeq 2$ and $\delta_f \simeq 2 \sim 3$, so that, with $< q >_{min} \simeq 2$, we expect the relevant flavoring rapidity scale to be $y_f \simeq 8 \sim 10$.

4.3 Effective intercept and Scale-Dependent Treatment

In order to be able to extend a Pomeron repesentation below the rapidity scale $y \sim y_f$, we propose the following **scale-dependent** scheme where we introduce a flavoring factor for each Pomeron propagator. Since each Pomeron exchange is always associated with energy variable s, (therefore a rapidity variable $y \equiv \log s$), we shall parametrize the Pomeron trajectory function as

$$\alpha_{eff}(t; y) \simeq 1 + \epsilon_{eff}(y) + \alpha' t, \tag{28}$$

where $\epsilon_{eff}(y)$ has the properties

- (i) $\epsilon_{eff} \simeq \epsilon_o \equiv \alpha_{\mathcal{P}}^{old} - 1 \simeq 0$ for $y << y_f$,
- (ii) $\epsilon_{eff} \simeq \epsilon \simeq 0.1$ for $y >> y_f$.

For instance, exchanging such an effective Pomeron leads to a contribution to the elastic cross section

$$T_{ab}(s,t) \propto s^{1+\epsilon_{eff}(y)+\alpha' t}. \tag{29}$$

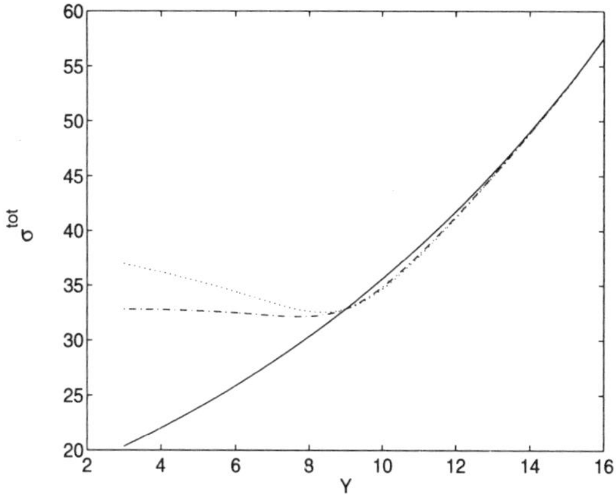

Figure 2: Effect of flavoring factor $R(s)$ when applied to a standard rising cross section: $\sigma^{cl} = \beta^2 s^\epsilon$, $\epsilon = 0.1$ and $\beta^2 = 16\, mb$, given by the solid curve. With $R(y)$ given by Eq. (37), the dashed-dotted curve has $\epsilon_o = 0$, $\lambda_f = 1$, and flavoring scale $y_f = 9$, and the dotted curve corresponds to $\epsilon_o = -0.04$.

This representation can now be extended down to the region $y \sim y_r$. We shall adopt a particularly convenient parametrization for $\epsilon_{eff}(y)$ in the next Section when we discuss phenomenological concerns.

To complete the story, we need also to account for the scale dependence of Pomeron residues. What we need is an "interpolating" formula between the high energy and low energy sets. Once a choice for $\epsilon_{eff}(y)$ has been made, it is easy to verify that a natural choice is simply $\beta_a^{eff}(y) = \beta_a e^{[\epsilon-\epsilon_{eff}(y)]y_f}$. It follows that the total contribution from a "flavored" Pomeron to a Pomeron amplitude is

$$T_{a,b}(y,t) = R(y)\, T^{cl}_{a,b}(y,t), \tag{30}$$

where $T^{cl}_{a,b}(y,t) \equiv \beta_a\beta_b e^{(1+\epsilon+\alpha'_{\mathcal{P}}t)\,y}$ is the amplitude according to a "high energy" description with a fixed Pomeron intercept, and $R(y) \equiv e^{-[\epsilon-\epsilon_{eff}(y)](y-y_f)}$ is a "flavoring" factor. In terms of $s = e^y$,

$$R(s) \equiv (\frac{s_f}{s})^{[\epsilon-\epsilon_{eff}(\log s)]} \tag{31}$$

where $s_f \equiv e^{y_f}$. (See Figure 2.)

This flavoring factor should be consistently applied as part of each "Pomeron propagator". With the normalization $R(\infty) = 1$, we can therefore leave the residues alone, once they

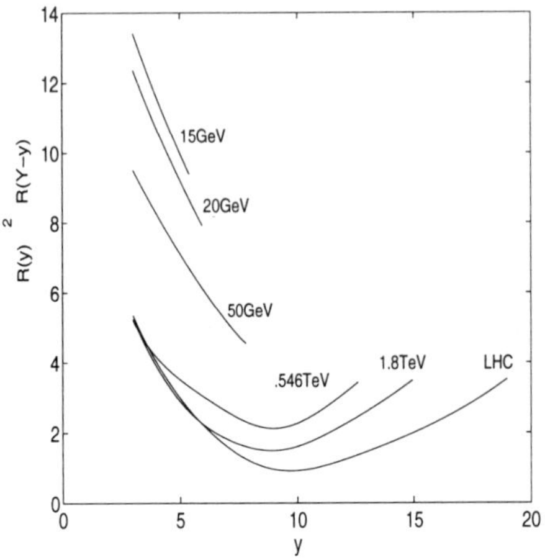

Figure 3: Renormalization factor due to flavoring alone, $Z_f(\xi;s) \equiv R^2(\xi^{-1})R(M^2)$, as a function of rapidity $y = \log \xi^{-1}$ for various fixed center of mass energies. These curves correspond to parameters used for the solid line in Figure 4.

have been determined by a "high energy" analysis. For instance, for the single-particle gap cross section before taking into account final-state screening, since there are three Pomeron propagators, one has for the renormalization factor $Z = R^2(y)R(y_m)$. It is instructive to plot in Figure 3 this combination as a function of either ξ or M^2 for various fixed values of Y.

5 Final Recipe

Having explained the notion of flavoring and its effects both on Pomeron intercept and on its residues, we need to go back to re-examine its effect for final-state screening. Since a Pomeron exchange enters as a Born term, i.e., the eikonal for either the elastic or the inelastic diffractive production, flavoring can easily be incorparated if we multiply both $\chi(y,b)$ and $\chi_d(y,b)$ by a flavoring factor $R(y)$. That is, if we adopt a generalized eikonal model for final-state screening, the desired screening factor becomes

$$S_f(\xi,t) = S_f^{(0)}(y,t;R(y)X(y),\mu_d) \tag{32}$$

where $S_f^{(0)}$ is given by Eq. (16). We have also explicitly exhibited the dependence on the maximal value of the flavored elastic eikonal, RX, and on the effectiveness parameter μ_d.

Let's now put all the necessary ingredients together and spell out the details for our proposed resolution to Dino's paradox. Our **final recipe** for the Pomeron contribution to single diffraction dissociation cross section is

$$\frac{d\sigma}{dtd\xi} = D_{a,\mathcal{P}}(\xi,t)\Sigma_{\mathcal{P}b}(M^2), \tag{33}$$

where the unitarized flux, $D_{a,\mathcal{P}}$, and the Pomeron-particle cross section, $\Sigma_{\mathcal{P}b}$, are given in terms of their respective classical expressions by $D_{a,\mathcal{P}}(\xi,t) \equiv Z_d(\xi,t)F^{cl}_{a,\mathcal{P}}(\xi,t)$ and $\Sigma_{\mathcal{P}b}(M^2) \equiv Z_m(M^2)\Sigma^{cl}_{\mathcal{P}b}(M^2)$. It follow that the total suppression factor is

$$Z(\xi,t;M^2) = Z_d(\xi,t)Z_m(M^2) = [S_f(\xi,t)R^2(\xi^{-1})]R(M^2), \tag{34}$$

with the screening factor given by Eq. (32) and the flavoring factor given by Eq. (31). Finally, we point out that the integral constraint for the unitarized flux, Eq. (5), when written in terms of these suppression factors, becomes

$$\int_{-\infty}^{0} dt \int_{0}^{1} d\xi S_f(\xi,t)R^2(\xi^{-1})F^{cl}_{a,\mathcal{P}}(\xi,t)\xi^{\epsilon} = \beta_a^{diff} < \beta_a(0), \tag{35}$$

where $F^{cl}_{a,\mathcal{P}}(\xi,t) \equiv (1/16\pi)\beta_a(t)^2 g(t)(\xi^{-1})^{2\alpha_{\mathcal{P}}(t)-1}$.

5.1 Phenomenological Parametrizations

Both the screening function and the flavoring function depend on the effective Pomeron intercept, and we shall adopt the following simple parametrization. The transition from $\alpha^{old}(0) = 1+\epsilon_o$ to $\alpha^{new}(0) = 1+\epsilon$ will occur over a rapidity range, $(y_f^{(1)}, y_f^{(2)})$. Let $y_f \equiv \frac{1}{2}(y_f^{(1)} + y_f^{(2)})$ and $\lambda_f^{-1} \equiv \frac{1}{2}(y_f^{(2)} - y_f^{(1)})$. Similarly, we also define $\bar\epsilon \equiv \frac{1}{2}(\epsilon+\epsilon_o)$ and $\Delta \equiv \frac{1}{2}(\epsilon-\epsilon_o)$. A convenient parametrization for ϵ_{eff} we shall adopt is

$$\epsilon_{eff}(y) = [\bar\epsilon + \Delta \tanh \lambda_f(y - \bar y_f)]. \tag{36}$$

The combination $[\epsilon - \epsilon_{eff}(y)]$ can be written as $(2\bar\epsilon)\,[1+(s/s_f)^{2\lambda_f}]^{-1}$ where $s_f = e^{y_f}$. Combining this with Eq. (31), we arrive at a simple parametrization for our flavoring function

$$R(s) \equiv \left(\frac{s_f}{s}\right)^{(2\bar\epsilon)\,[1+(\frac{s}{s_f})^{2\lambda_f}]^{-1}}. \tag{37}$$

With $\alpha^{old}_{\mathcal{P}} \simeq 1$, we have $\epsilon_o \simeq 0$, $\bar\epsilon \simeq \Delta \simeq \epsilon/2$, and we expect that $\lambda_f \simeq 1 \sim 2$ and $y_f \simeq 8 \sim 10$ are reasonable range for these parameters. [27]

To complete the specification, we need to provide a more phenomenological description for the final-state screening factor. First, we shall approximate the screening factor by an exponential in t:

$$S_f(y,t) \simeq S_f(y,0)e^{\Delta B_d(y)t} \tag{38}$$

where $S_f(y,0) = \{r\, x^{-r}\Gamma(x,r)\}^2$, with $x = \mu_d R(y)X(y)$ and $r = r(y)$. The width, $\Delta B_d(y)$, can be obtained by a corresponding substitution. Note that $S_f(y,0)$ depends on $B(y)$, $B_d(y)$, $X(y)$, and μ_d. Phenomenological studies allow us to approximate $B(y) \simeq b_0 + .25y$ and $B_d(y) \simeq b_d + .25y$, $b_d \simeq b_0/2 \sim 2.3\, GeV^{-2}$.

The only quantity left to be specified is the effectiveness parameter μ_d. Since the physics of final-state screening is that driven by a Pomeron with intercept greater than unity, the relevant rapidity scale is again y_f. Let us fix μ_d first by requiring that screening is small for $y < y_f$, *i.e.*, $S_f(y,0) \sim 1$ as one moves down in rapidity from y_f to y_r. Similarly, we expect screening to approach its full strength as one moves past the flavoring threshold y_f. We thus find it economical to parametrize

$$\mu_d(y) \simeq (\mu_0/2)\{1 + \tanh[\lambda_d(y - \bar{y}_f)]\}, \tag{39}$$

and we expect $\bar{y}_f \sim y_f$ and $\lambda_d \sim \lambda_f$. This completes the specification of our unitarization procedure.

5.2 Qualitative Discussion

Before attempting to "**use**" our recipe, it is important to gain a qualitative understanding of what one can expect. Note that the sole effect of the flavoring factor R is to "restore" a Pomeron contribution to that of a "bare Pomeron" with an intercept $\alpha_{\mathcal{P}}^{old} \simeq 1$. It follows that $R(y) \sim 1$ for $y >> y_f$ and $R(y) > 1$ for $y << y_f$. Similarly, since we expect final-state screening to increase as $\epsilon_{eff}(y) \sim \epsilon$, we therefore expect $S_f(y,t) < 1$ for $y >> y_f$ and $S_f(y,t) \to 1$ for $y << y_f$.

Let us examine the phase space region for a singly diffraction dissociation process from ISR to LHC. (See Figure 1.) It is customary to adopt $y \geq 3$ and $y_m \geq 0.4$ for a triple-Pmeron analysis, (corresponding to $\xi_{max} \sim 0.05$ and $\xi_{min} = 1.5/s$.) However, a triple-Pomeron analysis cannot be reliably used for $0 < y_m \leq y_r$. To be conservative, we adopt the same lower cutoff $y_{min} \simeq 3$ for both y and y_m. With $y_f \simeq 8 \sim 10$, the phase space is divided into the following four main regions of interest:

- (A) $y_{min} < y_m < y_f$, $y_{min} < y < y_f$:
  $\alpha_{eff}(y) \simeq 1$, $\alpha_{eff}(y_m) \simeq 1$, and $d\sigma^{sd}/dy \sim constant$.
- (B) $y_m < y_f$, $y_{min} < y < y_f$:
  $\alpha_{eff}(y) \simeq 1$, $\alpha_{eff}(y_m) \simeq 1 + \epsilon$, and $d\sigma^{sd}/dy \sim e^{\epsilon y_m} = (M^2)^\epsilon$.

- (C) $y_{min} < y_m < y_f$, $y > y_f$:
  $\alpha_{eff}(y) \simeq 1 + \epsilon$, $\alpha_{eff}(y_m) \simeq 1$, and $d\sigma^{sd}/dy \sim S_f(y,t)e^{2\epsilon y} = S_f(\xi,t)\xi^{-2\epsilon}$.
- (D) $y_m > y_f$, $y > y_f$:
  $\alpha_{eff}(y) \simeq 1 + \epsilon$, $\alpha_{eff}(y_m) \simeq 1 + \epsilon$, and $d\sigma^{sd}/dy \sim S_f(y,t)e^{2\epsilon y}e^{\epsilon y_m} = S_f(\xi,t)\xi^{-\epsilon}s^{\epsilon}$.

In region-(A), diffractive cross sections behave according to the "old-fashion" Triple-Pomeron formula, with $\alpha_{\mathcal{P}}(0) = 1$. Region-(B) is perhaps the best place where one has a clean chance of a direct detection of the $\epsilon > 0$ effect coming from the M^2-dependence at fixed $y < y_f$. One would normally also expect to be able to detect the $\epsilon > 0$ effect directly in Region-(C). However, in the large-y limit where the strength of final-state screening becomes important, the suppression from $S_f(y,t)$ will overcome the $e^{2\epsilon y}$ behavior. Region-(D) exits only if $Y > 2y_f$. With $y_f \simeq 8 \sim 10$, this region becomes interesting only as one moves from Tevatron to LHC energies.

Consider the two available energies at Tevatron. Moving along a fixed Y line, depending on the strength of screening, the cross section could be suppressed in the region-(C), and most of the contribution to σ^{sd} will come from region-(A) and region-(B). As one moves from $\sqrt{s} = 546\ GeV$ to $\sqrt{s} = 1800\ GeV$, the rapidity span in region-B decreases. In contrast, below Tevatron energies, one is most likely to be in region-(A). Even more strikingly, at these "low energies", a large portion of the kinematic region lies in regions where either y or y_m is near y_r or below. As a consequence, we would not be surprised if our triple-Pomeron formula turns out to lead to a σ^{sd} which is lower than what has been measured; other triple-Regge contributions should be included.

5.3 A Caricature of High Energy Diffractive Dissociation

The most important new parameter we have introduced for understanding high energy diffractive production is the flavoring scale, $s_f = e^{y_f}$. We have motivated by way of a simple model to show that a reasonable range for this scale is $y_f \simeq 8 \sim 10$. Quite independent of our estimate, it is possible to treat our proposed resolution phenomenologically and determine this flavoring scale from experimental data.

It should be clear that one is not attempting to carry out a full-blown phenomenological analysis here. To do that, one must properly incorporate other triple-Regge contributions, *e.g.*, the $\mathcal{PPR}$-term for the low-y_m region, the $\pi\pi\mathcal{P}$-term and/or the $\mathcal{RRP}$-term for the low-y region, etc., particularly for $\sqrt{s} \le \sqrt{s_f} \sim 100\ GeV$. What we hope to achieve is to provide a "caricuture" of the interesting physics involved in diffractive production at collider energies through our introduction of the screening and the flavoring factors. [27]

Let us begin by first examining what we should expect. Concentrate on the triple-Pomeron vertex $g(0)$ measured at high energies. Let us for the moment assume that it has also been meassured reliably at low energies, and let us denote it as $g^{low}(0)$. Our flavoring analysis indicates that these two couplings are related by

$$g(0) \simeq e^{-(\frac{3\epsilon y_f}{2})} g^{low}(0). \tag{40}$$

With $\epsilon \simeq 0.08 \sim 0.1$ and $y_f \simeq 8 \sim 10$, using the value $g^{low}(0) = 0.364 \pm 0.025\ mb^{\frac{1}{2}}$, [14] we expect a value of $0.12 \sim 0.18\ mb^{\frac{1}{2}}$. Denoting the overall multiplicative constant for our renormalized triple-Pomeron formula by K,

$$K \equiv \frac{\beta_a^2(0) g_{\mathcal{PPP}}(0) \beta_b(0)}{16\pi}. \tag{41}$$

With $\beta_p^2 \simeq 16\ mb$, we therefore expect K to lie between the range $.15 \sim .25\ mb^2$.

We begin testing our renormalized triple-Pomeron formula by first turning off the final-state screening, *i.e.*, setting $S_f = 1$. We determine the overall multiplicative constant K by normalizing the integrated σ^{sd} to the measured CDF $\sqrt{s} = 1800\ GeV$ value.[28] With $\epsilon = 0.1$, $\lambda_f = 1$, this is done for a series of values for $y_f = 7, 8, 9, 10$. We obtain respective values for $K = .24, 0.21, 0.18, 0.15$, consistent with our flavoring expectation. As a further check on the sensibility of these values for the flavoring scales, we find for the ratio $\rho \equiv \sigma^{sd}(546)/\sigma^{sd}(1800)$ the values 0.63, 0.65, 0.68, 0.72 respectively. This should be compared with the CDF result of 0.834.

Next we consider screening. Note that screening would increase our values for K, which would lead to large values for g. Since we have already obtained values for triple-Pomeron coupling which are of the correct order of magnitude, the only conclusion we can draw is that, at Tevatron, screening cannot be too large. With our parametrization, we find that screening is rather small at Tevatron energies, with $\mu_0 \simeq 0.0 \sim 0.2$ This comes as somewhat a surprise! Clearly, screening will become important eventually at higher energies. After flavoring, the amount of screening required at Tevatron is apparently greatly reduced.

Having shown that our renormalized triple-Pomeron formula does lead to sensible predictions for σ^{sd} at Tevatron, we can improve the fit by enhancing the PPR-term as well as RRP-terms which can become important. Instead of introducing a more involved phenomenological analysis, we simulate the desired low energy effect by having $\epsilon_o \simeq -0.06 \sim -0.08$. A remarkably good fit results with $\epsilon = 0.08 \sim 0.09$, $y_f = 9$ and $\mu_0 \simeq 0 \sim 0.2$. [27] This is shown in Figure 4. The ratio ρ ranges from $0.78 \sim 0.90$, which is quite reasonable. The prediction for σ^{sd} at LHC is $12.6 \sim 14.8\ md$.

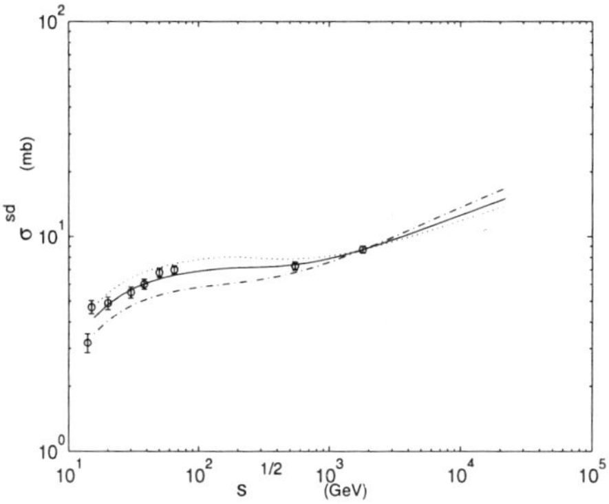

Figure 4: Fits to representative single diffraction dissociation cross sections from ISR to Tevatron.[5] The solid line corresponds to $\epsilon = 0.08$, $\epsilon_o = -0.07$, $\lambda_f = 1$, $y_f = 9$, with a small amount of final-state screening, $\mu_0 = 0.1$. The dotted and the dashed-dotted curves correspond to $\mu_0 = 0.2$ and $\mu_0 = 0$, i.e., no screening,respectively.

Our fit leads to a triple-Pomeron coupling in the range of

$$g_{\mathcal{PPP}}(0) \simeq .12 \sim .18 \; mb^{\frac{1}{2}}, \tag{42}$$

exactly as expected. Interestingly, the triple-Pomeron coupling quoted in Ref. [5] ($g(0) = 0.69 \, mb^{\frac{1}{2}}$) is actually a factor of 2 larger than the corresponding low energy value. [14] Note that this difference of a factor of 5 correlates almost precisely with the flux renormalization factor $N(s) \simeq 5$ at Tevatron energies.

We believe, with care, the physics of flavoring and final-state screening can be tested independent of the specific parametrizations we have proposed here. In particular, because our unitarized Pomeron flux approach retains factorization along the "missing mass" link, unambiguous predictions can be made for other processes involving rapidity gaps.

6 Predictions for Other Gap Cross Sections

For both double Pomeron exchange (DPE) and doubly diffractive (DD) processes, one is dealing with three rapidity variables which can become large. We will treat these two cases first before turning to more general situations.

Prediction for DPE Cross Sections:

For double Pomeron exchange (DPE), we are dealing with events with two large rapidity gaps. The final state configuration can be specified by five variables, t_1, t_2, ξ_1, ξ_2, and M^2. For t_1 and t_2 small, one again has a constraint, $\xi_1^{-1}M^2\xi_2^{-1} \simeq s$. Alternatively, we can work with rapidity variables, $y_1 \equiv \log(\xi_1^{-1})$, $y_2 \equiv \log(\xi_2^{-1})$, and $y_m \equiv \log M^2$, with $y_1 + y_m + y_2 \simeq Y = \log s$. The appropriate DPE differential cross section can be written down, with no new free parameter. Let us introduce a renormalization factor

$$\frac{d\sigma}{dy_1 dt_1 dy_2 dt_2} = Z_{DPE}(y_1, t_1, y_m, y_2, t_2)\frac{d\sigma^{classical}}{dy_1 dt_1 dy_2 dt_2}. \tag{43}$$

One immediately finds that, using Pomeron factorization for the missing mass variable,

$$Z_{DPE} = [S_f(y_1,t_1)R(y_1)^2]R(y_m)[S_f(y_2,t_2)R(y_2)^2] = Z_d(\xi_1,t_1)Z_m(M^2)Z_d(\xi_2,t_2). \tag{44}$$

Alternatively, we can express this cross section in terms of singly diffractive dissociation cross sections as

$$\frac{d\sigma_{ab}}{dy_1 dt_1 dy_2 dt_2} = \{\frac{d\sigma_{ab}}{dy_1 dt_1}\}\{R(y_m)\sigma_{ba}^{cl}(y_m)\}^{-1}\{\frac{d\sigma_{ab}}{dy_2 dt_2}\} \tag{45}$$

where $\sigma_{ba}^{cl}(y_m) = \beta_b\beta_a e^{\epsilon\, y_m}$. This **clean** prediction involves no new parameter, with the understanding that, when y_m is low, secondary terms must be added.

Prediction for DD Cross Sections:

For double diffraction dissociation (DD), there are two large missing mass variables, M_1^2, M_2^2, separated by one large rapidity gap, y, and its associate momentum transfer variable t. Again, for t small, we have the constraint $y_{m_1} + y_{m_2} + y \simeq Y$.

The classical differential DD cross section is $\Sigma_{a,\mathcal{P}}^{cl}(y_{m_1})\bar{F}_{\mathcal{P}}^{cl}(y,t)\ \Sigma_{b,\mathcal{P}}^{cl}(y_{m_2})$, where the classical "gap distribution" function is $\bar{F}_{\mathcal{P}}^{cl}(t,y) = (1/16\pi)e^{2(\epsilon+\alpha'_{\mathcal{P}}t)y}g_{\mathcal{PPP}}^2(t)$. After taking care of both flavoring and final-state screening, on obtains for the renormalization factor

$$Z_{DD}(y_{m_1}, y, t, y_{m_2}) = R(M_1^2)[\bar{S}_f(y,t)R(y)^2]R(M^2) \equiv Z_m(M_1^2)\bar{Z}_d(\xi,t)Z_m(M_2^2). \tag{46}$$

A new screening factor, $\bar{S}_f(y,t)$, has to be introduced because of the difference in the t-distribution associated with two factors of triple-Pomeron coupling. It can be obtained from $S_f(y,t)$ by replacing $B_d(y)$ by $\bar{B}(y) = \bar{b} + \alpha'_{\mathcal{P}}t$, where, by factorization, $\bar{b} = 2b_d - b_0$, ($\bar{b}$ is the t-slope associated with the triple-Pomeron coupling).

Alternatively, this cross section can again be expressed as a product of two single diffractive cross sections

$$\frac{d\sigma_{ab}}{dy_{m_1}dtdy_{m_2}} = \{\frac{d\sigma_{ab}}{dy_{m_1}dt}\}\{\bar{Z}_d(y,t)\sigma^{el}_{ab}(y,t)\}^{-1}\{\frac{d\sigma_{ab}}{dy_{m_2}dt}\}, \tag{47}$$

where $\sigma^{el}_{ab}(y,t) \equiv (1/16\pi)|\beta_a(t)\beta_b(t)e^{(\epsilon+\alpha'_{\mathcal{P}}t)y}|^2$. Other than the modification from S_f to $\bar{S}_f$, this prediction is again given uniquely in terms of the single diffraction dissociation cross sections.

Other Gap Cross Sections

We are now in the position to write down the general Pomeron contribution to the differential cross section with an arbitrary number of large rapidity gaps. For instance, generalizing the DPE process to an n-Pomeron exchange process, there will now be n large rapidity gaps, with $n-1$ short-range ordered missing mass distributions alternating between two gaps. The corresponding renormalization factor is

$$Z^{(n)}_{PE} = Z_d(\xi_1,t_1)Z_m(M_1^2)\bar{Z}_d(\xi_2,t_2)Z_m(M_2^2)\cdots\cdots Z_m(M^2_{(n-1)})Z_d(\xi_n,t_n). \tag{48}$$

Other generalizations are all straight forward. However, since these will unlikely be meanful phenomenologically in the near future, we shall not discuss them here. It is nevertheless interesting to point out that, if any cross section does become meaningful experimentally, flavoring would dictate that it is most likely the classical triple-Regge formulas with $\alpha_{\mathcal{P}}(0) \simeq 1$ that would be at work first.

7 Final Remarks

Let us briefly recapitulate what we have accomplished. Given Pomeron as a pole, the total Pomeron contribution to a singly diffractive dissociation cross section can in principle be expressed as

$$\frac{d\sigma}{dtd\xi} = [S_i(s,t)][D_{a,\mathcal{P}}(\xi,t)][\Sigma_{\mathcal{P}b}(M^2)], \tag{49}$$

$$D_{a,\mathcal{P}}(\xi,t) = S_f(\xi,t)F_{\mathcal{P}/a}(\xi,t) \tag{50}$$

- The first term, S_i, represents initial-state screening correction. We have demonstrated that, with a Pomeron intercept greater than unity and with a pole approximation for total and elastic cross sections remaining valid, initial-state absorption cannot be large. We therefore can justify setting $S_i \simeq 1$ at Tevatron energies.

- The first crucial step in our alternative resolution to the Dino's paradox lies in properly treating the final-state screening, $S_f(\xi, t)$. We have explained in an expanding disk setting why a final-state screening can set in relatively early when compared with that for elastic and total cross sections.

- We have stressed that the dynamics of a soft Pomeron in a non-perturbative QCD scheme requires taking into account the effect of "flavoring", the notion that the effective degrees of freedom for Pomeron is suppressed at low energies. As a consequence, we find that $F_{\mathcal{P}/a}(\xi,t) = R^2(\xi^{-1})F^{cl}_{\mathcal{P}/a}(\xi,t)$ and $\Sigma_{\mathcal{P}b}(M^2) = R(M^2)\Sigma^{cl}_{\mathcal{P}b}(M^2)$ where R is a "flavoring" factor.

It is perhaps worth contrasting what we have achieved with the flux renormalization scheme of Goulianos. [5] By construction, the normalization factor $N(s)$ is of the form which one would have obtained from an initial-state screening consideration. Although this breaks factorization, one might hope perhaps the scheme could be phenomenologically meaningful at Tevatron energies. Note that, for $\sqrt{s} > 22\ GeV$, the renormalization factor $N(s)$ has an approximately factorizable form: $N(s) \sim .25\ s^{2\epsilon} = .25\ (\xi^{-1})^{2\epsilon}(M^2)^{2\epsilon}$. it follows that the diffractive differential cross section remains in a factorized form:

$$\xi\frac{d\sigma^{sd}}{dtd\xi} \sim .25[\xi^{2\epsilon+1}F^{(0)}_{\mathcal{P}/p}(\xi,t)][(M^2)^{-2\epsilon}\sigma^{(0)}_{\mathcal{P}p}(M^2,t)]. \tag{51}$$

It can be shown that Eq. (51) leads to a diffractive cross section σ^{sd} which, up to $\log s$, is asymptotically constant. That is, the diffractive dissociation contribution no longer corresponds to the part of total cross sections represented by the Pomeron exchange. This is not in accord with the basic hypothesis of Pomeron dominance for total and elastic cross sections at Tevatron energies. [29]

Our final resolution shares certain common features with that proposed by Schlein. [8] At a fixed ξ, $Z_m(M^2) \simeq 1$ as $s \to \infty$ so that it is possible to identify our renormalization factor $Z_d(\xi,t) = S_f(\xi,t)R^2(\xi^{-1})$ with the flux damping factor $Z_S(\xi)$ of Schlein. In Ref. [8], it was emphasized that the behavior of $Z_S(\xi)$ can be separated into three regions. (i) (ξ_1, ξ_{max}) where $Z_S \simeq 1$, (ii) (ξ_2, ξ_1) where Z_S drops from 1 to 0.4 smoothly, and (iii) $(0, \xi_2)$ where $Z_S(\xi) \to 0$ rapidly as $\xi \to 0$. The boundaries of these regions are $\xi_1 \sim 0.015$ and $\xi_2 \sim 10^{-4}$. The first boundary ξ_1 can be identified with our energy scale, $s_r \sim \xi_1^{-1} \sim e^{y_r}$. If we identify the boundary between region-(ii) and region-(iii) with our flavoring scale y_f by $s_f^{-1} = e^{-y_f} = \xi_2$, one has $y_f \simeq 9$, which is consistent with our estimate. Since $S_f(\xi,t) \simeq 1$ for $\xi > s_f^{-1}$ and $R^2(\xi^{-1})$ drops from $R^2(1) \simeq s_f^{2\epsilon}$ to 1 at s_f, their $Z_S(\xi)$ behaves qualitatively like our renormalization factor. If one indeed makes this connection, what had originally been a mystery for the origin of the scale, ξ_2, can now be related to the non-perturbative dynamics of Pomeron flavoring. [30]

It should be stressed that our discussion depends crucially on the notion of soft Pomeron being a factorizable Regge pole. This notion has always been controversial. Introduced more than thirty years ago, Pomeron was identified as the leading Regge trajectory with quantum numbers of the vacuum with $\alpha(0) \simeq 1$ in order to account for the near constancy of the low energy hadronic total cross sections. However, as a Regge trajectory, it was unlike others which can be identified by the particles they interpolate. With the advent of QCD, the situation has improved, at least conceptually. Through large-N_c analyses and through other non-perturbative studies, it is natural to expect Regge trajectories in QCD as manifestations of "string-like" excitations for bound states and resonances of quarks and gluons due to their long-range confining forces. Whereas ordinary meson trajectories can be thought of as "open strings" interpolating $q\bar{q}$ bound states, Pomeron corresponds to a "closed string" configuration associated with glueballs. However, the difficulty of identification, presummably due to strong mixing with multi-quark states, has not helped the situation in practice. In a simplified one-dimensional multiperipheral realization of large-N QCD, the non-Abelian gauge nature nevertheless managed to re-emerge through its topological structure.[24,25]

The observation of "pole dominance" at collider energies has hastened the need to examine more closely various assumptions made for Regge hypothesis from a more fundamental viewpoint. It is our hope that by examining Dino's paradox carefully and by finding an alternative resolution to the problem without deviating drastically from accepted guiding principles for hadron dynamics, Pomeron can continued to be understood as a Regge pole in a non-perturbative QCD setting. The resolution for this paradox could therefore lead to a re-examination of other interesting questions from a firmer theoretical basis. For instance, to be able to relate quantities such as the Pomeron intercept to non-perturbative physics of color confinement represents a theoretical challenge of great importance.

Acknowledgments

I would like to thank K. Goulianos for first getting me interested in this problem during the Aspen Workshop on Non-perturbative QCD, June 1996. Intensive discussions with K. Goulianos, A. Capella, and A. Kaidalov at Rencontres de Moriond, March, 1997, have been extremely helpful. I am also grateful to P. Schlein for explaining to me details of their work and for his advice. I want to thank both K. Goulianos and P. Schlein for helping me to understand what I should or should not believe in various facets of diffractive data! Lastly, I really appreciate the help from K. Orginos for both numerical analysis and the preparation for the figures. This work is supported in part by the D.O.E. Grant #DE-FG02-91ER400688, Task A.

References

1. This is the first of a series of papers where we discuss the general subject of diffraction scattering at Tevatron energies. In this first article, (hep-ph/9706276), we concentrate on soft diffractive production. In a subsequent article, we shall discuss the relation between soft and hard Pomeron,[3] with particular emphasis on the "flavoring" mechanism leading to the picture of a "Heterotic Pomeron", (hep-ph/9302308).
2. For fits for soft Pomeron parameters, see: A. Donnachie and P. V. Landshoff, Phys. Lett. **B** 296 (1992) 227; J. R. Cuddel, K. Kang, and S. K. Kim, Phys. Lett., **B** 395 (1997) 311; R. J. M. Covolan, J. Montanha, and K. Goulianos, Phys. Lett., **B** 389 (1996) 176; M. Block, to be presented at VIIth Blois Workshop on Elastic and Diffractive Scattering, Seoul, June 10-14, 1997.
3. G. Ingelman and P. Schlein, Phys. Lett. **B** 296 (1992) 227.
4. We use $s_0 = 1\ GeV^2$ as the basic energy scale through out this paper. These "classical" expressions are: $F^{cl}_{\mathcal{P}/a}(\xi, t) = (1/16\pi)\beta_a(t)^2(\xi^{-1})^{2\alpha_{\mathcal{P}}(t)-1}$, and $\sigma^{cl}_{\mathcal{P}b}(M^2, t) = g(t)(M^2)^{\alpha_{\mathcal{P}}(0)-1}\beta_b(0)$. The triple-Pomeron coupling $g(t)$ can be in principle determined by data below $\sqrt{s} \leq 30\, GeV$ where cross sections are relatively insensitive to the choice of the Pomeron intercept value used. As far as I am aware of, the first attempt in using a "triple-Regge" analysis to interpret real experimental data was made by: D. Silverman and C-I Tan, "Relation Between the Multi-Regge Model and the Missing-Mass Sepctrum", Phys. Rev., **D2** (1970) 233. A formal postulation was first made by DeTar, Jones, Low, Tan, Weis, and Young, Phys. Rev. Letters, **26** (1971) 675. For other contemporaneous works, see: D. Horn and F. Zachariasen, "Hadron Physics at Very High Energies", (Benjamin, 1973).
5. K.Goulianos, Phys. Lett., **B** 358 (1995) 379.
6. With $\xi_{min}(s) = 1.5/s$, $\xi_{max} = 0.1$, $\epsilon \simeq .08$, $N(s) \equiv \int_{-\infty}^{0} dt \int_{\xi_{min}}^{\xi_{max}} F^{cl}_{\mathcal{P}/p}(\xi, t) d\xi \simeq .25 s^{2\epsilon} > 1$, for $\sqrt{s} \geq 22\ GeV$, and $N(s) \equiv 1$ for $\sqrt{s} \leq 22\ GeV$ where $\int_{-\infty}^{0} dt \int_{\xi_{min}}^{\xi_{max}} F^{cl}_{\mathcal{P}/p}(\xi, t) d\xi < 1$.
7. K. Goulianos, Proceedings of the 3rd Workshop on Small-x and Diffractive Physics, Argonne National Laboratory, USA, September 1996; K. Goulianos, Proceedings of the 5th International Workshop on Deep Inelastic Scattering and QCD (DIS-97), Chicago, USA, April 1997.
8. P. Schlein, Proceedings of the 3rd Workshop on Small-x and Diffractive Physics, Argonne National Laboratory, USA, September 1996; P. Schlein, Proceedings of the 5th International Workshop on Deep Inelastic Scattering and QCD (DIS-97), Chicago, USA, April 1997.
9. S. Erhan and P. Schlein, "Saturation of the Pomeron Flux Factor in the Proton by Damping Small Pomeron Momenta", to be submitted to Phys. Lett. For a difference in opinion, see Ref. [10]. See also: A. Brandt, et al., "Measurements of Single Diffraction at $\sqrt{s} = 630$ GeV; Implication for the Pomeron Flux Factor", to be submitted to Nucl. Phys.
10. K. Goulianos, "Comments on the Erhan-Schlein Model of Damping the Pomeron Flux

at Small-x", hep-ph/9704454.

11. When compared with the classical formula, our decomposition corresponds to having the triple-Regge coupling absorbed into the flux factor, D_k. For our subsequent discussion, the triple-Pomeron coulping $g(t)$ will be absorbed into the classical flux factor $F^{cl}_{\mathcal{P}/a}(\xi, t)$ and the Pomeron-particle cross section without $g(t)$ will be denoted by $\Sigma^{cl}_{\mathcal{P}b}(M^2)$.
12. C-I Tan, Proc. of 2nd International Conference on Elastic and Diffractive Scattering, ed. K. Goulianos, (Editions Frantieres, 1987) p347.
13. C-I Tan, Proc. of XIXth International Symposium on Multiparticle Dynamics, Arles, ed. D Schiff and J. Tran T. V. (Editions Frontieres, 1988) p361.
14. R. L. Cool, K. Goulianos, S. L. Segler, H. Sticker and S. N. White, Phys. Rev. Lett. **47** (1981) 701.
15. For diffraction dissociation, c must have same quantum numbers as the incident particle a. For definiteness, we discuss leading particle gap distribution relative to the incoming particle a with a positive, large rapidity, and assume $c = a$. For $p\bar{p}$ collision, actual diffractive cross section, σ^{sd}, is arrived at by taking into account contributions involving diffraction at both p and $\bar{p}$ vertices.
16. This mechanism has been recognized before. It was used in Refs. [13] and [17] to explain why a "maximal odderon" cannot be allowed in any hadronic scheme which admits an expanding-disk interpretation at high energies.
17. G. Finkelstein, H. M. Fried, K. Kang, and C-I Tan, Phys. Lett., **B** 232 (1989) 257.
18. W. Frazer, D. R. Snider and C-I Tan, Phys. Rev., **D8** (1973) 3180.
19. It is possible also to apply an eikonal model to study initial state screen for singly diffractive dissociation cross section. This indeed has been performed before, however, without taking final-state screening into account. [20] The fact that inelastic absorption takes place at small impact parameter, with surviving scattering allowed only at the edge of the expanding disk has also been noted there. Since our final-state absorption would remove all scattering within the disk, applying an eikonal initial-state absorption procedure becomes unnecessary.
20. E. Gotsman, E. M. Levin, and U. Maor, Phys. Rev., **D49** (1994) 4321.
21. T. K. Gaisser and C-I Tan, Phys. Rev., **D8** (1973) 3881.
22. C-I Tan, Proc. IX Rencontres de Moriond, Meribel, France (1974).
23. We include both $N\bar{N}$ and $c\bar{c}$ production as well as other effects. The effective degrees of freedom involved are "diquarks" and charm quarks respectively. For color counting, a baryon is considered as a bound state of a quark and diquark. See also: J. W. Dash and S. T. Jones, Phys. Lett., **B** 157 (1985) 229.
24. H. Lee, Phys. Rev. Lett., **30** (1973) 719; G. Veneziano, Phys. Lett., **B** 43 (1973) 314. See also, F. Low, Phys. Rev. **D** 12 (1975) 163.
25. A phenomnenological realization of QCD emphasizing the topological structure of large-N gauge theories is Dual Parton Model, (DPM). For a recent review, see: A. Capella, U. Sukhatme, C-I Tan, and J. Tran T. V., Physics Reports, **236** (1994) 225.
26. We do not include "semi-hard" production in the current treatment for soft Pomeron. Flavoring will indeed be the primary mechanism in our construction of a "Heterotic Pomeron". As mentioned in Ref. [1], this will be treated in a separate publication.

27. By choosing $\epsilon^{old} < 0$, it is possible to provide a global "average" description mimicking "secondary trajectory" contributions for various low energy regions. In Ref. [21], acceptable estimates are $\epsilon_o \simeq -0.11 \sim -0.5$.
28. The published CDF σ^{sd} values at $\sqrt{s} = 546$ and $1800\ GeV$ are $7.89 \pm 0.33\ mb$ and $9.46 \pm 0.44\ mb$ respectively. These values correspond to $\xi_{max} = 0.15$. We shall restrict $\xi < 0.05$ and t to be inside the extreme forwrad peak. For $\xi_{max} = 0.05$, we reduce these values by $\sim 8\%$ while maintaining their relative ratio of 0.834. For t to be within the extreme forward diffraction peak we scale down the ISR diffractive cross sections also by approximately 8%. This is appropriate for our determination of the triple-Pomeron coupling.
29. Ultimately, these two schemes can be differentiated by confronting experimental data. For our scheme, because of Pomeron pole dominance, it leads to a normalizable limiting gap distribution, $\rho_a(y;Y)$, *i.e.*, $\rho(y,t;Y) \to \rho_a(y,t)$. For $y < y_f << Y$, it is cut-off in y at least as fast as $\rho_a(y,t) \propto e^{-\epsilon y}$. In contrast, the flux renormalization scheme, Eq. (51), leads to a gap distribution of the form $\rho_a(y;Y) \propto e^{-2\epsilon Y} e^{\epsilon y}$, for $1 << y << Y$. Test of these two alternatives for either the normalization and the y-distribution can in principle be carried out by comparing data at two Tevatron energies by focusing on the region of fixed small t and $0.02 \geq \xi \geq 0.002$, $(y_{min} < y < y_f)$. Unfortunately, both behaviors seem to provide acceptable fits based on data presented in Ref. [10].
30. There are also several differences between our result and that of Schlein. First, our renormalization factor $Z(\xi, t; s)$ is t-dependent whereas Schlein's is not. At very small ξ, our suppression factor does not vanish as fast as that of Schlein: ($Z_S(\xi) \sim \xi$, whereas ours behaves as $\xi^{2\epsilon}$). Furthermore, since we have found that there is very little screening needed at Tevatron energies, our slow cutoff might not set in until much higher energies so that it could indeed be possible to observe the $\xi^{-2\epsilon}$ behavior for $d\sigma^{sd}/dydt$ in Region-(C) at Tevatron energies.

NON-ABELIAN EIKONAL FORMALISM IN QCD_4[1]

H. M. FRIED

Physics Department, Brown University

Providence, RI 02912 USA

ABSTRACT

Using a new technique for the extraction of leading-log energy dependence from the non-Abelian portion of quark-line interactions, hadronic scattering amplitudes may be represented in terms of functional integrals over gluonic material exchanged between scattering quarks. If that material is further represented by a condensed tube of gluonic flux (obtained by dimensional transmutation), these functional integrals will produce effective propagators corresponding to plasma oscillations, or "condensons", in this condensed gluonic material. These condensons yield a non-zero, non-tachyonic, finite, gauge-invariant contribution only when the gluonic flux tube corresponds to a "rigid string" of negligible thickness, so that this formalism provides a natural mechanism for a "dynamical string" in QCD4. Summation over a subset of such relevant condensons (at least in SU(2)) generates a form of the Donnachie-Landshoff Pomeron for the scattering amplitude. Other, non-condensate approaches are also available.

[1]Supported in part by DOE Grant # DE-FG02-91ER 40688

What can happen when a pair of quarks inside different hadrons have approached each other sufficiently so that they exchange a gluon? Immediately, that gluon splits into two, and then splits again and again into low-frequency gluons which interact with each other, converting that initial gluon into an entity which should not be described in terms of individual gluons, but rather in terms of a semi-classical, gluonic material, or flux. If that flux forms a tube with finite energy per unit length, one anticipates that such a flux tube may, in a non-scattering context, have something to do with confinement. Note that the gluonic flux in the tube need have nothing to do with "classical QCD", in the sense of satisfying classical field equations. In fact, these remarks are valid in a dimensional transmutation (DT) context[1], as long as the tubes are understood to be "rigid", corresponding to flux pointing in the same spatial direction at every point of the tube. For technical reasons, such "rigidity" is needed in the DT analysis; and it is also reminiscent of the "thin needles" proposed a decade ago by Preparata.[2]

What is the radius of such a tube? DT (as calculated in the gluonic sector) does not say; but if the "plasma oscillation" excitations, or "condensons" - which correspond to the relevant degrees of freedom in the tube's condensed flux - are to be emitted and absorbed by the scattering quarks, then the condenson propagator can contribute only if the flux tube has a vanishingly small radius, i.e., if the tube can be thought of as a rigid string. And the only portion of the condenson propagator which does contribute is that part which is non-tachyonic, effectively gauge invariant, and finite. A summation over the simplest subset of such relevant condensons can be performed for SU(2), and produces[3] a scattering amplitude of Donnachie-Landshoff form.[4]

The above scenario is not the only one possible. Starting from the non-Abelian version of the eikonal approximation of quark-line interactions, one can display qq scattering amplitudes in terms of functional integrations (FIs) over the gluonic field generated by and exchanged between the scattering quarks. In these representations, which contain in principle, the full gluonic interactions necessary for overall gauge-invariance of the amplitude, the non-Abelian eikonal analysis[5] allows one to extract, in a straightforward, functional way - using a new principle called "contiguity" - the leading-log energy dependence of the FI integrands, leaving the momentum-transfer dependence buried in the gluonic fields and hadronic wave functions. The challenge is then to extract the proper s- and t-dependence of the amplitude, by rescaling the FI variables, by creating a differential equation in s and t

directly from the FI, or by some other, as yet unappreciated method. At present, the only results known to the author are those of the DT analysis described above.

It will perhaps be most efficient to devote the remaining paragraphs of this note to the new, non-Abelian, quark-line, eikonal representation which underlies the entire discussion. The exact scattering amplitude can be cast in the form of appropriate FIs over Green's functions corresponding to the scattering quarks, and in the mass-shell-amputated limit, each quark will produce an ordered-exponential (OE) factor of form

$$\left(\exp\left[-ig\int_{-\infty}^{+\infty} ds\, p_\mu\, A_\mu^a\,(z - sp)\,\lambda_a\right]\right)_+ \tag{1}$$

which is to be linked, by appropriate FI, to that OE of the other quark. The difficulty is, and has long been, that these factors are OE's, rather than ordinary exponentials, and the relatively simple structure of Abelian FIs is of no avail. Rather, the only obvious, available method for evaluation has been that of perturbative expansion[6] in the coupling, which is not exactly what one would prefer to do. For simplicity, and to restrict the discussion to its most essential features, let us consider the SU(2) problem where interactions between the exchanged "rho-mesons" are neglected, and the only non-Abelian aspect treated is the coupling of two such OEs, which could be those of nucleons; here, the coupling is via a FI over a Gaussian (in A) weighting factor. The FI is trivial in the corresponding Abelian problem, but impossible to avaluate exactly for our SU(2) example.

The first step is to rewrite each OE in terms of a pair of other FIs (over proper-time-like variables), with N' a normalization constant,

$$()_+ = N' \int d[\alpha] \int d[u] \left(e^{i\int_{-\infty}^{+\infty} ds\lambda_a u_a(s)}\right)_+ \cdot e^{i\int_{-\infty}^{+\infty} ds\alpha_a(s)[u_a(s)-gp_\mu A_\mu^a(z-sp)]}, \tag{2}$$

so that the $\int d[A]$ FI is Abelian, and may be immediately performed. Neglecting all self-energy, vertex, $\cdots$ structure along each quark line, one is then left with the formidable result,

$$\begin{aligned} R &= N' \int d[\alpha] \int d[u]\, e^{i\int_{-\infty}^{+\infty} ds\, u\cdot\alpha} \left(e^{i\int_{\infty}^{+\infty} ds\lambda^I\cdot u}\right)_+ \\ &\cdot\ N' \int d[\beta] \int d[v]\, e^{i\int_{-\infty}^{+\infty} dtv\cdot\beta} \left(e^{i\int_{-\infty}^{+\infty} dt\lambda^{II}\cdot v}\right)_+ \end{aligned} \tag{3}$$

$$\cdot \quad \exp\left[i\int_{-\infty}^{+\infty}\int ds\, dt\, \alpha_a(s)\, Q_{ab}(s,t)\, \beta_b(t)\right],$$

where in this simplest case $Q_{a,b}$ is the isotopic propagator built out of the ρ-meson propagator,

$$Q_{ab}(s\cdot t) = p_{\mu 1}\, p_{\nu_2}\, \Delta_{c\mu\nu}\,(z_1 - sp_1 - z_2 + sp_2)\,,$$

and $(z_{\mu 1}, p_{\mu 1})$ and $(z_{\mu 2}, p_{\mu 2})$ are the CM spatial quark 4-momenta of the scattering problem, in the limit of zero momentum transfer.

How then does one evaluate or estimate the R of (3)? For this we have found it very worthwhile[7] to ask: "What would Schwinger do?" And the answer is that he would introduce extra parameter and source dependence, in an effort to find a "differential characterization" - that is, a differential equation in parameter and source variables - for R. There are various ways of doing this, but perhaps the simplest is to replace the R of (3) by the $R(s;\eta)$ defined as

$$\begin{aligned} R(s;\eta) &= N'\int d[\alpha]\int d[u]\, e^{i\int_{-\infty}^{+\infty}\alpha\cdot u}\left(e^{i\int_{-\infty}^{s}\lambda^I\cdot u}\right)\cdot \\ &\cdot\quad N'\int d[\beta]\int d[v]\, e^{i\int_{-\infty}^{+\infty}\beta\cdot v}\left(s^{i\int_{-\infty}^{+\infty}dt\lambda^{II}\cdot v}\right)_+ \\ &\cdot\quad e^{i\int_{-\infty}^{+\infty}\int \alpha_a Q_{ab}\beta_b}\cdot e^{i\int_{-\infty}^{+\infty}dt' v_b(t')\eta_b(t')}\,, \end{aligned} \tag{4}$$

where s is a parameter and $\eta(s')$ is a source; clearly, $R = R(s;\eta)|_{s\to\infty,\eta\to 0}$. By calculating $\partial R/\partial s$, and a sequence of integrations-by-parts, one can obtain a differential equation linear in $\partial/\partial s$ and $\delta/\delta\eta$, whose obvious solution is the OE (in s'):

$$R(s;\eta) = \left(\exp\left[i\int_{-\infty}^{s}ds'\int_{-\infty}^{+\infty}dt'\lambda_a^I\, Q_{ab}(s',t')\left[\eta_b(t') + \Lambda_b^{II}\left(t'|\frac{\delta}{\delta\eta}\right)\right]\right]\right)_{+(s')}, \tag{5}$$

where

$$\Lambda_b^{II}(t|iv) = \left(e^{i\int_t^\infty \lambda^{II}\cdot v}\right)_+ \lambda_b^{II}\left(e^{-i\int_t^\infty \lambda^{II}\cdot v}\right)_-,$$

and is an ordered solution of the integral equation

$$\Lambda_b^{II}\left(t|\frac{\delta}{\delta\eta}\right) = \lambda_b^{II} - 2if_{bcd}\int_t^\infty dt'\,\frac{\delta}{\delta\eta_c(t')}\,\Lambda_d^{II}\left(t'|\frac{\delta}{\delta\eta}\right). \tag{6}$$

Note that the double-ordered R of (3) still contains two sets of enmeshed orderings; but that they are defined and can be discussed separately.

Solutions to (5) and (6) are not trivial; but there is a new "contiguity" technique[5] which extracts that part of R which is clearly going to be exponentiated (in the form of an ordinary exponential, of course), and which when expanded in powers of the coupling corresponds to extracting the leading-log energy dependence of all relevant Feynman graphs. Further, for our SU(2) problem, all such terms can be summed without difficulty, and with $R = \exp[i\chi]$, generate for the eikonal χ of this problem the quantity

$$\chi = -\frac{g^2}{2\pi} \left(\sigma^I \cdot \sigma^{II}\right) K_0(\mu b) \left\{1 - \frac{4}{3}(1 - e^L) + \frac{2}{3} L\, e^L\right\}, \qquad (7)$$

where $$L = \frac{g^2}{\pi^2} \ln\left(E/m\right)\ K_0(\mu b)\,, \qquad \text{with} \qquad b = z_{\perp 1} - z_{\perp 2}\,,$$

the impact parameter of the problem. Here, the Pauli σ-matrices for each particle have replaced the Gell-Mann λ_a of SU(3).

This eikonal is real, because no (multiperipheral) mechanism for meson emission has been included; but it does provide a high-energy limit for this limited SU(2) problem, after a hiatus of more than two decades.

The techniques used above are applicable to other, more modern and relevant situations, such as the construction of full, QCD scattering amplitudes. In particular, the combination of a flux-string description of the gluons with our $R(s;\eta)$-contiguity technique for the quarks holds out the promise of being able to begin with standard QCD and to end with realistic scattering amplitudes. Of course, the use of non-interacting condensons can only be an approximation, if a useful one; and an effort should be made to extract those terms from the full, gluonic stew which will be necessary for the production of a "triple Pomeron", as well as Cheng-Wu-type multiperipheral processes necessary for absolute unitarity restrictions, although the latter could conceivably be of a non-Froissart form. One even might expect that an appropriate, "static" sum over all non-interacting condensons will be quite relevant to confinement. In essence, this is neither phenomenology, nor QCD lattice magic, but rather an example of analytic, non-perturbative QFT relevant to experiment.

References

[1] H. M. Fried, *Phys. Rev.* **D** (1992).

[2] See, for example, G. Preparata, *Nuovo Cimento* **A103** (1990) 1073.

[3] H. M. Fried and Y. Gabellini, "An Eikonal Calculus", in preparation.

[4] See, for example A. Donnachie, Proc. of the Workshop on Quantum Infrared Physics at the American University of Paris, June 1994; published by World Scientific, Singapore (1995).

[5] H. M. Fried and Y. Gabellini, "Non-Abelian Eikonals", *Phys. Rev.* **D** to be published.

[6] See, for example, H. Cheng and T.T. Wu, "Expanding Protons: Scattering at High Energies", The MIT Press, Cambridge (USA), 198.

[7] In other words, one does not have to be a Schwinger in order to make progress; but, rather, one only need do as he, in other contexts, did.

FISHING IN THE UPPER ROARING FORK VALLEY

Peter Carruthers

Aspen and Tucson

September, 1993

A Guide to Assured Success and Prestige in Which

Secrets Are Revealed for the First Time in Print

Worthy of the Perusal of Most Anglers

PREFACE

After more than thirty years fishing the beautiful streams around Aspen, Colorado, I finally decided to share my admittedly limited knowledge in the form of this small book. Even though other fishermen may know more, there is presently no little guide with information on where to go, how to get there, and what to expect and do once you are there.

The presentation is organized into two main parts, namely a description of the tributary streams, and the big water from Aspen down to Upper Woody Creek. The first is written with beginners in mind, although the description of the characteristics of each of the headwater streams will be of interest to the visitor. We give special attention to access. Without knowing how to get to good water, visitors will be handicapped.

This is not a book on technique nor is it an instruction manual in fly tying. It is more a book on strategy, and what to expect when you go to various places, and how to get to those places. Althought the focus of the book is specific to the Aspen area, there should be much of interest to fishermen throughout the Rocky Mountains. In order to lighten the monotony of relentless information we have included various tales to amuse the reader.

In addition we include some information on the side benefits of trout fishing around Aspen. These include an awareness of bird life, mushrooms, wildflowers and such. They certainly increase the pleasure of fishing in beautiful places. On a good day you will see an orchid, pick raspberries in the stream bed, catch a few trout, and pick some good mushrooms. This sort of experience creates the peaceful illusion of self-sufficiency, so unsuited to current life.

I actually believe that much of the urge to fish is animal in nature. In daily life these primordial tendencies are suppressed. That is one reason I come down hard on the excesses of the "catch and release" crowd. I also am pained to see new-style fishermen who seem to feel that the sport supplies a patina of modern age sophistication to their curriculum vita, without really understanding what it is all about. They are the most susceptible to politically correct thinking.

By all means take a child fishing. But let him keep it, take it home to eat and share. Then he or she will better understand a part of life that is disappearing.

In this text we discuss safety here and there. Experienced anglers will hopefully not need this advice. Among the hazards is eating a misidentified mushroom. Although the ones listed here are unique and unmistakable to the experienced fungophile, beginners can make sickening mistakes. It is important to have a good book and still better, an experienced friend.

With regard to our primitive maps, note that solid lines denote streams, closed solid lines lakes or reservoirs, dashed lines roads, and dotted lines trails.

Acknowledgements always present difficulties. There are many friends with whom I shared much in Aspen over the past 32 years, and who directly or indirectly contributed to this book. Some are lampooned herein, I hope with a light touch. Through this time the Aspen Center for Physics has provided a perch for me and my families. The beauty of the valley and mountains and the inspiration of the music make life worth living.

INTRODUCTION

This unparalleled treatise originates from a pamphlet I wrote at the Aspen Center for Physics sometime in the early 70's. Many of my colleagues were interested in learning about trout fishing, but didn't know what to get, where to go, nor what to do when they got there. Besides, they did not want to spend much money. Physicists are misers except when in a good restaurant.

To deal with this crisis I composed a brief guide for beginners, emphasizing small streams around Aspen. I promised a sequel "Deep Trout" treating the big waters of the Roaring Fork (now called "quality" water). Here it is. Hopefully the regional focus and emphasis on simplicity will make this little book useful to both novices and experts alike. Much of the strategy and most of the fishing problems are common to rivers and streams throughout the western mountains. Most important, we suggest where to go, access, and discuss in detail what to expect in various locations

To make things clear, my original audience was, and still is, interested in putting food on the table in front of an admiring family. These anglers, who view trout as food in addition to other things, hold in contempt the politically correct view that trout must be caught and released. As in all PC systems, the violators are supposed to be ashamed, and to keep quiet. Just as we had a short rod craze, followed by a long rod fad, and presently a knotless leader rule, there will be backlash (pun intended), as sure as short skirts follow long skirts.

Of course the quality waters are presently catch and release. Personally I prefer to eat trout from the small tributaries, anyway.

In recent years a new type of fisherman has arrived on the scene, to partake of the new prestige sport called fly fishing. I started to call them "yuppies" but my wife noted that perhaps young people don't know the expression. You can tell their influence by looking at the current slick fishing magazines. In fact all you need to know is in Ernest Schwiebert's "Matching the Hatch" and Herter's wonderful catalogues, alas no longer with us. To get in the right mood scan Milford "Stanley" Poltroon's "How to Fish Good". By not taking yourself too seriously you can use a size 12 hook, instead of size 20, without feeling guilty.

This new breed supports the guide industry and no doubt buys a lot of fancy equipment. Anybody who gets a copy of this (no doubt by illegal photocopy) won't need guides. The novice angler only needs to learn how to cast, tie a few knots, and follow the author's advice.

Not all the advocates of catch and release are amateurs or ideologues. In fact many are professional fishermen and guides. My only plea is that you learn how to remove the hook carefully. I see too many deformed fish, many with multiple open wounds. What is the point of torturing the trout? While the "spared" trout dies from its wounds the hypocrite fisherman eats a restaurant imitation grown in a concrete bathtub in Idaho. There are plenty of trout around here if you only take your share.

Our real problem as fishermen is to preserve the rivers, lakes and wetlands. Despite the diligence of organizations like Trout Unlimited, it is very hard for a group composed of part time members with limited funding to fight off those interests which would destroy our waterways. I was once president of a chapter of TU in Los Alamos, New Mexico. In Ithaca, New York I belonged to an active chapter. We worked hard to repair eroded streams, stocked trout, and all the usual things. However disillusionment comes fast in volunteer conservation organizations. For example, we discovered that raw sewage from the old-folks home drained into Cayuga Lake. When I brought in samples the county refused to accept them. After two years of county denial the problem was partially fixed. And we all know that it is impossible to get up early enough to prevent the Corps of Engineers from straightening a lovely meandering meadow stream. How else can they get a quick runoff that creates floods and justifies a new dam?

There are many ways to get discouraged in this business. Here in Aspen, on the "west slope" of the Rocky Mountains, we see a slow but study increase in the diversion of water in our rivers to feed the thirsty and politically invincible city of Denver.

To temper these frustrations I recommend fishing in the beautiful streams of the upper Roaring Fork. These waters are an unknown wonder, so luckily most visiting fishermen crowd onto the Frying Pan, which has drive up service and lots of company. (Anyway, Ruedi Reservoir covered up my favorite holes.)

The many feeder streams run down from the mountains through beautiful forests and canyons. Each has a different character. I will give a brief description of the characteristics of most of these. However there is so much water to cover that I must leave many things for the reader to discover.

We then focus on the "big water", that part of the Fork which begins in Aspen and goes several miles down to upper Woody Creek bridge. Along the way the river becomes increasingly large and powerful as it goes though Aspen, picking up the flow of Hunter Creek, Castle Creek and and Maroon Creek. As this swift torrent moves over the beautiful red, white and green rocks, there are days that you would not guess how many big fish are there hiding in the shade of the canyon. I break the description into well defined segments, each a possible afternoon fishing trip. Perhaps the details that I offer will help you get along without a guide (expensive and not always informed). However if you are a beginner it is essential to learn the basics of correct casting. There are special problems casting in mountain rivers that I will address later.

Note that what I call "big water" is currently called "quality water" or "gold medal water". I cannot bring myself to use these tasteless expressions. Are they selling something to drink, along with some flour? Do they imply that the other fishing opportunities are of lower quality? What brain dead committee chose these obnoxious expressions?

In order to temper the boredom of where to, what to, when to etc. I have inserted little vignettes from true life experience.

PROLOGUE

As in a close relationship, rivers begin small and develop in ways determined by means not known to me. The Roaring Fork begins near Independence Pass, breathtaking to the easterner and westerner alike. From Independence Lake and others the flow trickles past the formerly beautiful city of Independence. In the 1960's the image of the old mining town was truly impressive. Now it is gone through neglect.

Many other tributaries feed the river. Each has its own charm.

About twelve miles east of Aspen, Lincoln Creek (formerly known as the South Branch) joins the other branch of the river. If you drive up Lincoln Gulch on the formerly rough road there is beautiful scenery, canyons, raspberries, blue grouse, and ptarmigan at high elevations.

Twenty miles downstream limpid water flows quickly over the river bottom of red, white and green rocks Granite boulders up to thirty feet wide divert the flow, creating falls and deep holes. The Roaring Fork River (so named by the Indians) still roars despite sizeable diversions to the east slope. In Aspen and several miles downstream the "Fork" runs through a canyon, whose walls block much of the view of the overconstruction in and around the city of Aspen. Orange algae cover the boulders, providing a contrast to the dark blue green color of the spruce and the willow green of the cottonwood. Dippers build nests, lemmings slink by; an occasional deer is to be seen. This great river seems little known compared to it's famous neighbor, the Frying Pan. That is fine with me.

First I will discuss fishing in the headwaters. The fish are usually not large, but there are are plenty and the beauty of these valleys must be seen to be believed. Then I will describe the big waters, the central part of the "Fork" occupying only a few miles of the main river basin, which must have a linear dimension.of fifty miles. On the east it begins at Hallam Lake in the city of Aspen and ends at the upper Woody Creek Bridge. After that the river spans out into shallow fast riffles, in large, although many deep holes are to be found.

Everybody is familiar with the Aspen of the rich and famous, the fancy hotels and the multitude of private jets at the airport. (By the way the flight path goes right over the Upper Woody Creek part of the river). In the summer one can go to the Aspen Music Festival situated on a beautiful meadow. Nearby stand Walter Paepke's famous Aspen Institute for Humanistic Studies.and the Aspen Center for Physics. In this womb-like place one can think of Samuel Johnson's novel "Rasselas, A Prince of Happy Valley", Once inside you do not care what happens outside. Often a lecturer at the Physics Center has to pause while Beethoven finishes a chord in the music tent next door.

All around are stately mountains, tall grass, sage and flickering Aspen leaves. In the evening the radiant brilliant sun is blinding as it disappears into the night. I love this Aspen, which is such a combination of beauty, culture and decadence.

It is necessary to have a nostalgia section. In 1962, when my love affair with Aspen began, only Main Street was paved. The gravel in the back streets had an especially sharp and nice smell. Independence Pass road was unimproved gravel and probably safer than now. I fished away peaceful afternoons below town, where it was totally silent, save two trains going by on the old railroad bed. The engineer always waved, happy to see another person. Now this area is jammed with bicycles, dogs, joggers and horses. It was better the other way.

As everybody knows, the character of Aspen has changed a lot. The originals are either dead or gone. I remember first arriving in 1962, going to Sabbatini's, then a sporting goods store on Mill Street. I could not afford the hip boots; they were an astronomical $42. Mrs. Sabbatini knew that I was a scientist, and always made me add up the bill in my head. In those days I still bought flies. One day I made a mistake, which she corrected. Phil Wright, a noted fisherman, ran the Country Store, which has gone through many hands and names over the years. Phil had an incredible supply of fly fishing tackle. He was rarely there since he was out fishing. His son would take me to the basement where I could select among the hundreds of necks. Nowadays it is impossible to find any serious fly-tying material in Aspen. Bring your own, or go to Basalt, which has emerged as a serious fishing center due to the fame of the Frying Pan. In those days there were no washing machines in condos; even house rentals had these precious instruments off limits. Everybody went to a single laundromat, the Whale, where locals, musicians, physicists, executives, writers were all thrown together.

The visual slum now known as the Airport Business Center was formerly part of a giant field of sage and grass, a cow pasture, in short. To get to the river you needed to check the location of the bull, in order to make sure that you could go the quarter mile to the river. Then you looked down on a lovely meadow and a great stretch of the Fork, rather than two settling ponds and the whirrings of the sewage plant motors.

I remember returning to Aspen from Rome in the summer of 1968, only to be appalled at the massive outbreak of condominium construction.

Despite many problems, Aspen is hard to ruin. My impression is that more people are shopping, and fewer going outdoors. Walk half a mile, and you will be in the National Forest. Thank goodness for that.

But what Aspen really means to me after thirty summers is the Roaring Fork itself and all the beautiful streams feeding it, and the forests and meadows through which these rivers run.

TABLE OF CONTENTS

Chapter 1

Fishing in the Headwaters of the Roaring Fork

1.1 Introduction

This chapter is designed for newcomers to fly fishing. Hopefully there are some passages of interest to the more experienced fisherperson.

This work is based on thirty one years of diligent piscatorial research in the Aspen area. By following the enclosed advice you can astound yourself as well as your friends by putting trout on the table, projecting a perhaps undeserved image of competence in this most difficult art form. Aimed at the inexperienced and incompetent, the text is laced with wry wit not to mistaken for condescension. In the edition of the early 70's there were references to nude swimmers in the grotto area up the pass. Alas those days are gone for now.

Here we present valuable advice, worthy of the perusal of most anglers, for beginners or others intent on filling their bellies with the least expense and effort.

We shall mostly be concerned with streams rather than lakes. Mountain lakes contain notoriously fickle trout. Remember that if you can see them they can see you. One way to deal with the large distances in such lakes is to use spinning gear, a bubble float partly filled with water to provide throwing weight, attached two or three feet above the fly (such as a gold ribbed hare's ear nymph). Lost Man Lake, Maroon Lake and Grizzly Reservoir are accessible places to fish. The trout are stocked in most of these places.

Fly fishing is by far the most reliable way to catch trout in streams except when the water is muddy or exceptionally cold. As shown by Professor Zachariasen in his piscatorial researches, no particular skill is required to be a successful meat fisherman using this technique. The main question is where to go and, what to do when you get there.

- **Interlude: Portrait of a Meat Fisherman**

  In recent times it has become politically incorrect to kill a trout for the unspeakable purpose of eating it.

  Little would you guess that Dr. Sydney Meshkov, formerly of the National Bureau of Standards and now at the Supercollider in Dallas sallies forth whenever hunger strikes to collect dinner with the same lack of guilt occasioned by picking wild mushrooms. Syd is a cheerful outgoing man, not some bestial monster, who fishes to keep and to eat. Not given to affectation, he usually goes to small waters or lakes and reservoirs, using spinning gear loaded with a bubble and one of my gold-ribbed hare's ears. Unperturbed by the guilt that is so fashionable nowadays, he does his job very well with no pretensions to elegance. I doubt if he has heard about "no-hackle" flies.

  Catch and release is imagined by some to improve the quality of fishing. Sometimes it is appropriate in extreme examples of too much pressure (as in the case of the Frying Pan) in order to preserve or revitalize overfished or polluted waters, but often there is little point. Although a very few killer fishermen are possible hazards if they kept what they caught, I do not accept the current dogma any more than the story that milk, butter and eggs are bad for you. After thirty years of falling into the quality water, I suggest that the fishing is worse nowadays, not better. Twenty years ago you could keep six daily over 12 inches. The next experiment was to allow one to keep trout less than twelve inches, on the amateur genetic theory that this would produce more large trout. Now it is all catch and release. This produces dumb trout with multiple head wounds. Anyone who thinks that the Fork can be fished out is an egomaniac, given that so much of the water is unfishable.

  I recall seeing a breed of fisherman in the '60's that does not seem to exist anymore. Often in early evening a grizzled sun-rumpled old man would hit the Fork. He (and they) fished in jeans and threw a big wet fly tied to a heavy leader strait downstream. After landing a couple of nice trout they would head home for dinner.

  Catch and release has all the appeal of safe sex. There is a genuine killer instinct in real fishermen, and its satisfaction in moderation is not sinful. As a boy I looked with awe at the tubs of giant bluegills caught by my grandfather. I understood two things: there was food and these fish really fight.

  So, a pox on those self-righteous jingle-jangle fishermen who have been brainwashed out of their manhood.

  Eat and enjoy. Long live Syd.

1.2 What to Expect

I am going to describe fishing in typical small streams 10-20 feet in width. These streams are mostly shallow, say a foot deep, with occasional holes. The choicer (deeper and protected) spots are seized by the larger trout. When feeding, the trout are often in very shallow water. Because of this they are easily spooked; you should always be about twenty feet downstream from your quarry. For similar reasons it it better to cast upstream even if you are using a wet fly. The reason of course is that the trout has his nose pointed into the current waiting for some food to drift by. Because the water is shallow you might as well use a wet or dry fly just under the surface. If you use a fly with some white, such as any variant of the Rio Grand King, you can usually follow the action. In deep holes I recommend sinking a nymph to the bottom.

Frequently the trout are all over the stream in shallow places that are almost impossible to fish with bait or hardware. Cast near rocks (submerged or above the surface), near undercut banks, where currents run into deeper pools, and many other places that experience will teach you. In small streams trout can often be seen feeding, for example in the shallow tail of a pool. Here you can pick off your trout one by one although some tricky current problems can arise.

When you come to a typical "long" pool (say 50 feet long) with rapids running into deep water, past some rocks, then through a riffle getting shallower towards the tail, begin by casting in the shallow water in the tail and gradually ease upstream. This way you will not scare away additional potential victims, who are looking upstream for their own prey. In a stream like Hunter Creek it is common to catch ten trout in one such pool. A certain stealth and delicacy of presentation will help. Do not kick the rocks! Believe it or not, some people do.

Pause: How to Fall Gracefully

Unless you can walk on water, you have to walk on rocks. Because of a fast current, algae, your bad judgement or a perfidious rock that is unstable, you will fall. Usually it is cool and refreshing but sometimes you can banged up be. As you fall push your rod and reel high to protect them. Your survivors might be able to use them. Try to fall vertically. Of course you will have put your wallet or other valuables high in a tight pocket. Often you will want to drain your boots. In late summer the water usually is not too cold and you could wear old jogging shoes. This simplifies transporting boots but is tough on your ankles and shoes. You will fall anyway. Keep your creel fastened so that your hard won quarry do not float downstream in the event of a fall.

- **Interlude: Truit Bleu**

  One of the best things to do in Aspen is to convince Maurice Thuillier, chef for many years at the Aspen Alps, to close the restaurant for an evening to host a "cooking class" In our case this typically involves about a dozen inhabitants of the Aspen Center for Physics. Syd Meshkov, Aspen habitue, devoted meat fisherman and fungophile, uses his enthusiasm to organize these classes. The class consists of four hours listening to Maurice' unique repartee while he delivers course after course of truly fine French cuisine cooked with commentary before your eyes. It is washed down by quite acceptable wine.

  Presently Maurice is pretending to be retired but we fans hope to get him back for more classes. Over the years this has become a ritual. Some take notes at the beginning.

This silly activity doesn't last long. A lot is learned and forgotten; the main lesson is that any unit of clarified butter, say two tablespoons, is to be interpreted as a cup.

Although Maurice collects his own wild mushrooms, he does not fish. One night I proposed to bring in some live trout to be subjected to the brutal torture and death by boiling known as truit bleu. The appointed day arrived and my wife Lucy, Fred Zachariasen and I set off to the smaller Fork east of Aspen. Often you can knock off ten trout in an hour in these waters. In the back of our four wheel drive was a large tub full of water to keep the victims alive. Soon the water was sloshing all over as we bumped along Lincoln Gulch. Fred's job was to stand at a respectful distance downstream to net the copious trout catch and quickly get them in the tub. (Fred is a Cal-Tech physicist; you can tell that we are close friends because of his constant invective about my various faults. His fishing fantasy is to use dynamite).

As luck would have it, a strong storm was coming over the mountains from the east. Soon it was pouring down rain, and all trout had taken refuge under the rocks. There it was: my reputation in danger. Trout were so scarce that as they slowly appeared we gave them names. The first was Marvin, then Melvin finally appeared. About five I dumped my helpers and went out again and finally caught one more who went unnamed since Fred, the name executive, was not there. So much for fun.

Exhausted, I showed up with three fish in a tub, still alive. Since they were a little larger than I had planned, things worked well. Of course the trout wiggled out from Maurice's hands, hopping across the kitchen with Maurice in crouched pursuit. For this crime they were ripped open and doused in boiling fluid of some kind. After all, fleeing from the authorities is a very serious crime.

Among the various lessons to learn from this tale is that a strong storm from the east can really kill the fishing in around Aspen.

Casting in the headwater streams presents problems not apparent when you learn classic casting techniques. The banks are dense with willows and such to grab your fly. Spruce trees are the worst. Master the roll cast. Very often the best place is overhung with brush so that precise side casts are essential. Often it is smart to look behind with every backcast to make sure that you are safely away from the brush. Keep your backcast high (I often extend my entire casting arm). Even with care, expect to lose several flies per trip. Brook trout have a special strike technique that can remove your fly. First of all they sometimes strike in a circular motion that enhances their apparent weight. This strike is very quick and savage. Secondly, if your fly is dangling in the current downstream the suddenness of the strike can break off the fly. A little bit of loose line helps solve these problems.

1.3 Equipment

If you don't have a waterproof hat you will be sorry. Please no white or red. Make sure that it won't blow off in a gust of wind. Khaki is the best color.

Don't forget bug repellent, sun lotion and a few first aid items.

- **Rod, reel and line.** It is not too expensive to set up a basic fly tying outfit. As with other sports, your fun and effectiveness usually increase with the quality of your equipment. I would recommend a 71/2 foot graphite rod, say 2-3 weight, with a good floating double tapered line to match. For a while weight forward lines were popular. The story is that the macho fisherman can "shoot" the line farther (farther than one can responsibly respond to strikes). Anyway this is not the Florida Keys. Stick to one of the many good double tapers made nowadays. I am told that the Cortland firm makes a decent graphite rod for about $60 and a rod, reel and line ensemble for about $140.

  A longer rod helps you get tangled in the jungle around you. I personally prefer a 1-weight 6 1/2 foot rod for small stream fishing but beginners will find a 7 or 7 1/2 foot rod easier to handle. Fiberglass pack rods that I know do not cast well; I have not tried the graphite. Orvis and Thomas and Thomas make attractive but expensive graphite pack rods that I would like to try. Inexpensive and functional reels are readily available. Cortland and Pfleuger are strong and reliable.

  After an apprentice period with a cane rod I began as a boy with a Sears-Roebuck telescopic steel rod. I must have been very strong. Later came 7' and 9' Shakespeare Wonder Rods. These glass rods caught many fish of many species. The 7' was my workhorse for years. I scoffed at anglers with fancy equipment while they sneered at me for fishing with such junk. Finally I learned about graphite and fine reels. Once corrupted,..... I love the feel of bamboo but I fear that they could not take all the buffeting that a rod gets bushwhacking in the high country around Aspen.

  If you have a friend who is a fishing addict, perhaps you can borrow the basic equipment. Just don't fall on the rod.

  The list price for a standard big name fly line has increased to about $40. To outsmart the rascals use your line for many years. Eventually it will develop a lot of little cracks and you will need a new one. I keep my eyes open for big discount sales and stock up. I prefer double tapers to weight forward (you can use both ends) and also prefer simple white to the incandescent pinks that you see. Perhaps these do not scare the fish. I was scared to try. While it is very useful to balance the reel and rod sizes, sometimes you will find that a heavier-than-recommended line is helpful, especially for casting at short distances.

- **Leaders.** Standard 7 1/2' with a 5X to 6X tippet is fine. I far prefer knotted tapered leaders (especially Dan Bailey Platlon) but they are hard to find as the new fad pushes knotless ones. In my view the latter do not cast as well. A good knotted leader is composed of segments of different stiffness, a feature not to be found in the currently favored ones. If you get into beaver ponds or mountain lakes you will do best with as long a leader that you can handle tapered to 6-7X. If your casting is insecure, use around nine feet with a thin tippet, say 6X). Getting the cast started is a chore but you need to make long casts anyway. Watch out for willows, and around mountain lakes, the steep rock bank rising behind you, eager to break off the point of your hook.

- **Polaroid Glasses.** Sometimes these are very helpful in cutting the glare from surface reflection. I have a pair of prescription polaroids. Often I leave them at home. There is only so much room in my vest.

- **Hip boots.** These are essential in small streams. Strangely most of the few remaining fishing stores in Aspen push expensive and uncomfortable chest high waders. They are needed for the deep placid waters of the Frying Pan and some deeper parts of the big water section of the Fork. To me they are as uncomfortable as panty hose seem to be. Waders are an encumbrance in the tributaries. Some of the cheaper brands will split if a beaver snag looks at them. Get a good pair and of course a patching kit. They should last several years unless you store them all folded up in a hot place.

  Whatever you get to keep the cold water at bay, make sure that a good fit with the chosen socks is a reality. Sometimes you have to walk a long way.

- **Others:**

  - Canvas Creel: Get a good one. If you put fish in it, wet it occasionally and put it in the stream if you stop to rest. It also helps to break off tips of willow branches to insulate one fish from another. When on a long trek, especially hiking in the hot afternoon sun, I usually clean the fish before returning.
  - Landing net Optional. Pull the fish on the bank and crush him with your great weight. Put half his body in the creel before dehooking since trout are strong and slippery. However they are acrobatic and can wiggle straight up and out if you are not on guard. If you hook a big trout you will really need a net. Modern technology has yet to learn how to carry a net that does not catch on everything. There is the folded type designed like a switchblade but it is ugly.
  - Fly box, floatant, knife, thin slices of lead, scissors, cord, whiskey and water battle. Warm beer is essential. A colored ribbon to mark your return route is useful.You can put it in the river under a rock. Scissors on a wound up string attached to a spring are not essential but make life easier. Goop to float dry flies.
  - Paper and plastic bags to put finds such as mushrooms.
  - Carry an emergency whistle. Many of the recommended places are chosen so that people do not ordinarily go there. If you are alone and become immobilized you will need the whistle.
  - Fishing vest to put all this in. Maybe a friend has an old grimy one in the closet. Ask the wife.

- **Interlude: How to Survive a Trip to the Fishing Store**

  As you may have found, a visit to the store to pick up a leader can turn into a massive shopping spree. Perhaps there is something like new car spray that erodes those nerves in the brain that control such impulses. Be warned that your resistance can dissolve in seconds as you see all the new doohickeys that appear every year.

  Remember that the proprietors are capitalists. Some of them actually know what they are doing. In any case be prepared to ignore their sad eyes when they hear about your current equipment. As with all styles, the politically correct views change year by year. Probably your current setup is fine. If you wait long enough your equipment will again be fashionable, even a valuable antique. You may have the impression that the rod and reel are the most expensive components of your setup, but you could be wrong. Now there are so many overpriced, mostly unnecessary cute little gadgets that you need to be wary.

  Take a walk in the brisk air, look at your check book, recall your credit card balances as you wait for the hormone flush to wear off. Above all, do not go to a bar to think it over.

1.4 Short List of Locally Successful Flies

In the small tributaries very few varieties are needed. For most trips the wet Rio Grande King (tied down white calf tail "wings") in sizes 12- 16 are all that I need. If you want variety try a Renegade (14-16) or a small Humpy. Mosquitoes and black ants work. The Ginger Quill, dark and light Cahill are good in small sizes. Usually almost anything works in these little rivers. (But not always!). If you wander into beaver ponds or lakes size 20 midges are advised. Tie them yourself. After the first 100 they will begin to look OK although not professional. Very little material is needed for these flies.

Remember the all-purpose miracle flies: the Adams (size 12-18, usually 14-16) and the gold-ribbed hare's ear nymph (size 14-18, usually). Have a few put in your grave to puzzle archaeologists.

The following ingredients allow you to tie such basics as the Rio Grand King, Adams, ant, mosquito and muskrat: Golden pheasant nape. Black chenille. Hard white and dark brown rooster hackle. Deer hair and some Black and yellow chenille. Tiny black hackle for the gnat. Moose mane with mixed black and white fibers. Medium grade small grizzly neck for the Adams and mosquito.

A few miscellaneous wet flies will come in useful. Make sure that you have a variety of sizes and colors.

1.5 Safety

This is a less trivial issue than you might think. Many of the target fishing areas are in the back country, and reached on potentially dangerous four wheel drive roads. Sometimes the wading is trivial, at other times terrifying in high fast water. Just today I got tired crossing

Castle Creek (lots of snowmelt, even in August). Too tired to recross, I went on all fours up a slippery slope grasping at willows.

The banks are often nothing but a jungle of brush, almost impossible to get through. One good thing: there is no poison ivy or poison oak here. Beaver cause a lot of trouble beyond your first surprise of sinking 18 inches into some muddy goo. Beaver snags are the number one cause of wader death. Beaver works include sneaky channels, pits and so forth.

Ever scratch your cornea on a twig? If you do go home instantly to lie in misery on the couch for two days. Stop at the emergency room on the way to get some cortisone drops.

Some streams (as Lincoln Creek) have gorge sections with beautiful deep pools and big trout (big means a foot long). As you go upstream you will encounter pools too deep to wade, with granite cliffs on both side. A glance shows it to be barely possible to scale the cliff and come down the other side, so up you go. About ten feet up you stop since it seems impossible to go further or to go back down. Stay put until the shaking stops and you will get out. It has occurred to me simply to jump in the pool.

I should advise that you always have a partner aware of your existence. However I do not follow my own advice.

Do not count on the water being safe to drink. Carry your own, unless you are high in the mountains with delicious snow-melt all around. An even more dangerous feature of fishing in general is its disruptive force in a relationship. Although women like to get rid of you once in a while the idea that a fish inspires greater enthusiasm than they is something you must conceal at all costs. Speak of your blood pressure, get a prescription from a (male) doctor. I remember a Sunday thirty years ago when my first wife threw a tantrum upon learning that I was going fishing. I silently thought: this is the end, and soon it was. There is danger in both directions.

Take a police whistle if you get in real trouble. After today I will get one. For a change you might be glad to see the police.

Chapter 2

Local Tributaries and Their Characteristics

2.1 Introduction

In this long chapter I give synoptic information on most of the interesting small tributaries of the Fork in the Aspen area. In many cases I lack detailed experience; the valley is very large. Yet some of the information, especially access, which actually changes with time, is hard to get. Your planning will be aided very much by the excellent "Pitkin County Open Space and Trails Map". It has a perfect scale, except that it it occupies a rectangle of 31 48745 by 60 52391 inches The regular local topographical maps are useful but do not have the detailed contemporary information that you need.

2.2 Lincoln Creek

Leaving Aspen we drive east up the Independence Pass Road. Besides the fantastic scenery we are distracted by heavy traffic and cumbersome "camping" trailers. Ahead we see Green Mountain, guarding the entrance to Lincoln Gulch, and New York Peak, on the south side of the gulch. We pass Weller Lake Campground and. the Grottos, encountering the first real turnoff on the right on the Lincoln Creek Road.

The road is rough but passable in an ordinary passenger car. After reaching the campground (1/4 mile) we encounter the stream; the dirt road follows Lincoln Creek for the 6-7 miles to Grizzly Reservoir, which stores water for diversion to the east slope. Many individual camp sites are placed along the stream, and there is Portal Campground at the reservoir. This is the trail head for a pretty seven mile walk up to Grizzly Lake, which has a good trout population.

Now back to the Lincoln Creek campground, to begin our trip.

Lincoln Creek is a beautiful stream in a beautiful valley. A dense spruce forest banks the stream until we get close to the reservoir. Typical birds include Ruby Crowned Kinglets, Gray Jays (to steal your scrambled eggs), Stellar's Jays, Robins, an occasional Pine Grosbeak, Chickadees, Nuthatches, Flickers and Williamson's Sapsuckers. The altitude here is about 10,000 feet. Although I have caught fish along the entire river, I usually drive about one mile before starting to fish. However, there are nice undercut banks all along the stream. These often harbor rainbows. Look for the mushroom Hydnum imbricatum late in the season. A popular name for this fungus is Hedgehog mushroom. It is copious in August and often rather good, though some are bitter. The top looks like dried leather and the underside not gilled but composed of small fleshy "teeth".

Shortly you will see a deep pool right by the road, with a central boulder. Although much fished, this pool often has something of interest. However the main point is that, having fished left around a corner, you enter a beautiful canyon, first with falls and small pools. Reach over the lip of the fall to fish in the pockets above. With care you will not be seen. Continuing upstream you will find deep pools with perhaps 20 substantial trout, usually nymphing. These fish spook easily. After you have caught one, all the rest disappear. Proceeding through these deep spots is difficult, since there are many impassable rocks and cliffs. Be careful. Don't forget to eat the raspberries.

After going up a few steep slopes, notice the lovely water-carved rocks on the right. This is a favorite picnic spot (too favorite). Also the upstream stretch up to the New York Gulch cutoff are among my favorite beats. Wading is not hard although flooding can deposit debris on the bank. Walking through the forest is the usual nuisance but not a problem. Keep your eyes open for mushrooms.

Almost every place is good unless a team of bait fishermen has attacked recently. The fish are a mix of brooks and rainbows, the brooks mostly 8 inches or less and the rainbows rarely over 12 inches. Usually a wet Rio Grand King (size 12-14) is all you need. Many dry flies work. A match the hatch situation is rare.

About half way to the reservoir the road to New York Gulch turns right, but is now blocked off. The ugly gash in the mountain to the southeast carries the waters of New York Creek and Brooklyn Creek over to Grizzly Reservoir. These streams are too small to encourage trout fantasies.

The stream continues through varying terrain, with less of a drop than in the lower part. There are many beaver works and little channels full of trout.

If you get to the reservoir, fish the inlet. Beyond Portal Campground it is a nice walk up to the old mining town of Ruby (about four miles). Here the little stream is not worth fishing. However, there several mountain lakes nearby.

2.3 The Upper Fork

I have never found this part of the watershed to be interesting. Most of it is shallow water with few holes to sustain larger trout. Perhaps I am missing something

In late July there are good hatches of the famous steinpilz mushroom in the spruce forests near Lost Man, provided there is enough rain. The mountain trails in this area can provide chanterelle, also in Willow Creek, which is now closed. The chanterelle make the best sauces I know for most meat dishes.

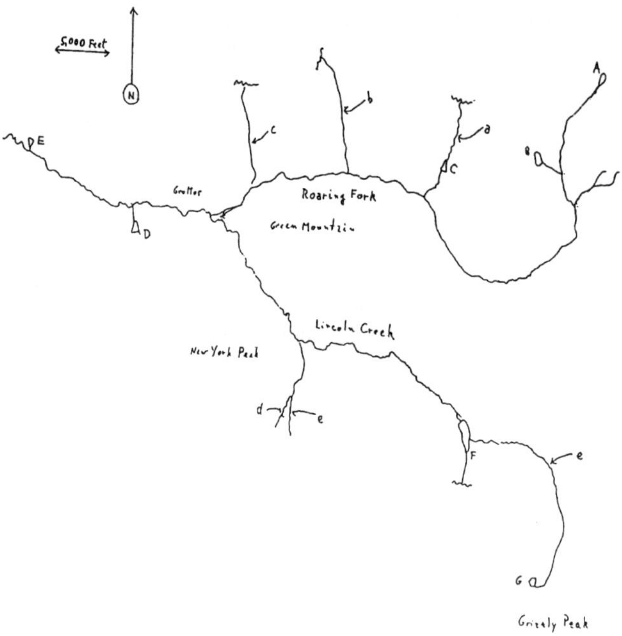

Figure 2.1: Lincoln Creek, the Upper Roaring Fork and the Grotto area are shown. The tributaries of these are not recommended for fishing. Grizzly reservoir and Lost Man Lake are good for spinning and bait fishing. Tagerts Lake is private. There are many nice walks of any length in this region. Your family may wish to enjoy one of these as you fish: (A) Independence Lake, (B) Linkins Lake, (C) Lost Man Reservoir, (D) Weller Lake, (E) Tagerts Lake, (F) Grizzly Reservior, (G) Grizzly Lake; (a) Lost Man Creek, (b) Coleman Creek, (c) Ptarmigan Creek, (d) New York Creek, (e) Brooklyn Creek.

2.4 The Grotto Area

After the Upper Fork joins with Lincoln Creek, the nature of the river changes. There are many deep holes much prized by swimmers and sunbathers. Around 1970 they were often nude. No longer politically correct. Rapids connect the pools. The river is very close (too close) to the road. After various tries, I concluded that a combination of too many people and too many bait fishermen make this area unattractive. Sometimes in the evening trout can be caught here. Hatchery trucks replenish the supply frequently. Below Weller Lake the river plunges down steep cascades as another level of the pass is reached. Again the nature of the river changes.

2.5 Weller Lake to Difficult Campground

Following the cascade, the river is fast, with many pools, side runs, and beaver diversions. At one point it splits into two parts. In general a little exploration here will help locate many places rarely visited. The stream is typically 20 feet across, sometimes more. There are some wading problems in fast or deep spots, but nothing serious unless you are careless. Do not do as I do, i. e. false cast as you wade. This is one of the most attractive and peaceful places for a fisherman to meditate.

This continues for about three miles to the campground. Along the way is Tagerts Lake (private). This is the only private land in this stretch of river. As you drive up the pass you will see lots of rock fields and distant beaver ponds. I usually fish fairly far up, where an Aspen forest blocks the view of the valley. There is a sequence of pulloffs, with shade. I usually stop at the next to last before the road narrows over rock cliffs, with quaint guard rocks. If you get to this spot you have gone too far. This pulloff is quite large and is easily recognizable.

In order to get down to the river you have to walk through a steep slope with dense undergrowth through an open Aspen forest. You will come to a vestigial road that parallels the river. Be sure to tie a piece of something bright on a branch so that you know where to go back up. Otherwise you can spend a lot of time looking for your car. Keep your eyes open when you return as there is a second (worse) trail on the edge of the ravine through which the river runs. By no means go up the rock cliff as I once did.

The trail downstream is easy to walk and very pleasant. You can walk several minutes and then go down to the stream and fish back up. In the fall watch closely for the delicious Honey Mushroom (Armillariella mellea). It grows in bunches on tree roots and dead wood. With a little experience it is easy to recognize and is quite plentiful. Study your mushroom book carefully. Do not collect if you cannot see the dead wood upon which it feeds.

Fishing is straightforward although you will see that the trout are getting more picky as you move downstream. Cover as much water as you can. On a sunny day you should get several nice rainbows 8-12" and some brook trout. Yesterday they liked a size 16 Adams. The Physics Center fishing consortium assembled the name "fish market" since dinner could be got in a couple of hours. This name was deduced from Einstein's general theory of relativity.

Although I have fished this area since the 60's without seeing a posted sign and having never been scowled at, I was surprised when, about three years ago a man on a horse arrived on the little road to announce that I was trespassing. As I complained he explained that there was a new owner who didn't want anybody on his property. I hope that this is not the case. As in the sad case of Hunter Creek, we see trolls using their property rights to prevent access to wild places.

The Weller Lake campground is a good place to stay. The river is not special for fishing; a short pleasant walk takes you to the lake. There is decent fishing there. This area is distinguished by the famous Steinpilze mushrooms in season (late July after a lot of rain). It grows in association with blue spruce along the banks of the stream. (This fungus has international fame. Some names are Boletus Edulus (Latin), Porcini (Italian), Cep (French), Borovik (Russian). In wet weather you will also find the "Milk Cap" mushroom Lactarius deliciosus in the meadows interspersed among the spruce groves.

2.6 Difficult Creek

It has been a while since I fished here. It is accessed from the campground. A pack trail goes uphill (steep) and soon the stream separates from the trail. The terrain is rough. Cutthroats exist in the upper reaches.

2.7 Gravel Pits

Now we are getting close to Aspen (about seven miles east of Aspen) and houses are beginning to appear. As we pass the entrance to Difficult Campground there suddenly appears a 35 MPH sign. From here to town the police like to give speeding tickets. One day I was with Steve Berry as he coasted his antique jeep down the pass. We were nabbed for doing 42 in the 35 zone. Steve offered the jeep to the policeman if he could make it go faster than 35. He got off with an equipment citation.

Soon on the left you will see a sign for the Wildwood School. Turn in and go a short way to the river. There is a good parking lot. Downstream there was formerly a gravel pit. You can still see some evidence of it. Going downstream you will encounter a nice long pool, where trout (all three local species) are usually rising. Fish this carefully, changing flies as needed. These trout are not big but somewhat picky. Today they spurned my Rio Grande King and my Adams, but took a size 16 Ginger Quill with abandon.

Continue downstream where you will see a big sign announcing your entry to the sacred "gold water". That means catch and release. You are supposed to return the trout until you reach MacFarlane Creek upstream from the sign. This section is pleasant but quite ordinary. Why it is gold is a mystery to me. Upstream from the pool the river is about forty feet across, somewhat rapid, with boulders and all sorts of little pools and runs. It is easy to wade. The easy access and availability of sport makes this an attractive place close to Aspen.

As indicated, you should watch for speed traps on the road between Difficult Campground and the Aspen.

2.8 North Star Preserve

As the Roaring Fork approaches Aspen it enters a giant meadow, in which it cuts a sinuous path. Cottonwoods decorate the path of the river.as it meanders slowly over a muddy bottom. Although there are some large brown trout here, the insect supply seems less than that of other waters in the valley.

Until recently the Preserve was preserved by being a ranch. It is now protected from the development that is degrading so much of the valley.

To fish here long casts and stealth are required. Catch and release.

2.9 Ute Park

Head south on Original Street off Cooper (between Spring and West End Streets), then turn left onto Ute Avenue. Just after passing a cluster of new overpriced and overcrowded mansions you will find a parking lot on the left. Before you lies a paved portion of the Rio Grande Trail, which leads to the Aspen Club. As soon as you come to the river scramble down to fish upstream. The wading is tricky, since the rocks are just the wrong size and can be slippery from algae late in the season. Here you will find rainbows up to ten inches and an occasional brown (usually near boulders) at the side up to 15". As you go under the bridge connecting the north side to the Aspen Club, the rocks go away and the stream spreads out a bit. There are fish in the runs coming out of a deep long pool. The pool has many trout, somewhat wary because of fishing pressure. It is worth close investigation. You might have to match the hatch. Above the falls feeding the pool go upstream to a long slow pool. The water is very calm and not deep although it is easy to get in over your boot tops. There are small brook trout and occasional browns here. Long casts and a fine leader are necessary for success. You must wade very quietly.

Returning to the Aspen Club you will see a wall of giant houses, on the north bank. (In this morning's paper I saw an ad for 2.5 acres upstream from this. The asking price is only $ 2,950,00). The sight is rather depressing. One also sees condos with one foot in the river bed. We can pray that the Fork will roar once again.

2.10 Hunter Creek

To the north Red Mountain looms over the city of Aspen. To the east is Smuggler Mountain, laden with mine dumps and not yet built up like Red Mountain. Between is a classic hanging valley, created in glacial days, standing about 1,000 feet over the city. The stream cascades down the steep slope, joining the main Fork close to Hallam Lake.

Having reached the top you will find a cheerful stream bubbling along through meadows and open Aspen forests. There is a dirt road running parallel but not too close to the stream. The road has been closed so you have to walk or ride a mountain bike. Having once spent the afternoon digging my four-wheel drive out of a mudhole I don't mind the walk. The meadows exhibit another problem during rainy weather. The first car gets through but the soil turns into slime, leaving a second to flounder in the mud.

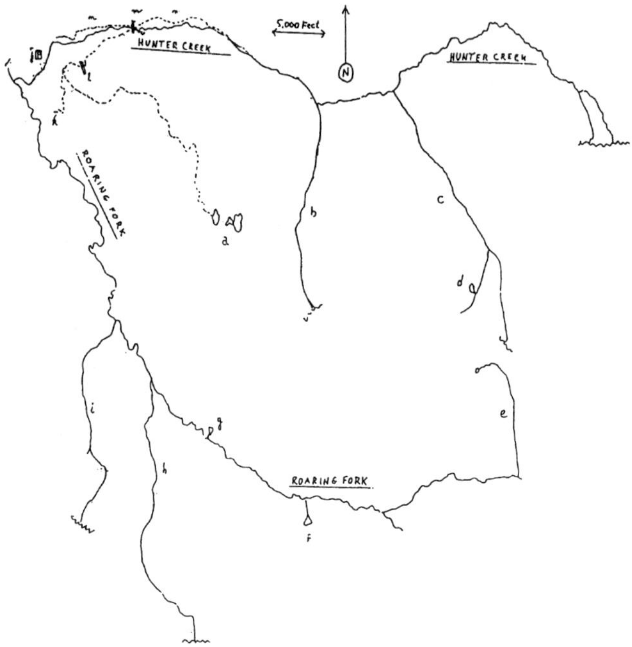

Figure 2.2: Hiking access to Hunter Creek is better shown on the map "Hallam Lake to Slaughterhouse Bridge". The four wheel drive access up Smuggler Mountain is shown, but not very well. Note that the gate (l) past which you have to walk to bridge (m). For further details consult the text: (a) Warren Lakes, (b) No Name Creek, (c) Midway Creek, (d) Midway Lake, (e) Coleman Creek, (f) Weller Lake, (g) Tagerts Lake, (h) Difficult Creek, (i) McFarlane Creek, (j) Red Mountain Parking, (k) Smuggler Road, (l) Forest Service Gate, (m) Reservoir Bridge, (n) Dirt Road (Blocked).

The valley is essentially east-west. On a late summer afternoon the tall yellow-green meadows glow with a brilliant light that I have not seen elsewhere.

Poem: Hunter Creek

In the afternoon the western sun
Illuminates translucent meadows

Shimmering tall green grass
In the hanging valley

There is no color like this
Finches bounce in the sky

And the wind bends the grass
As gold and green distort the mind

Happiness is to camp
Under dark spruce trees

While the night rain falls
While the campfire burns slowly

Getting to the best fishing in Hunter Creek takes time. A thorough visit will use up much of a day.

Access 1: Lone Pine. Go north on Mill Street, then turn left at the T junction and immediately right on Lone Pine. Now immediately left to a parking lot to the east of Hunter Longhouse, a condominium building. The trailhead is at the east end of the lot. If you come to the extensive Hunter Creek Apartment complex you have gone too far.

Access 2: Red Mountain. Go north Mill and turn left on Red Mountain Road at the T junction. Grind up the steep hill and take the right fork of a Y-junction onto Hunter Creek Drive. At numbers 0255 and 0209 you will see a blue sign turn left and park. Come onto road and walk uphill to the left. Ignore the "private" sign; it does not mean no trespassing. Soon you will spot another sign and the trail which goes up and crosses to the trail. You save some elevation and time by using this option.

Red Mountain residents are especially unfriendly to visitors parking on their streets. Vigilantes will have your car towed away if you are so audacious as to park it anywhere but the official lot. These people have created a visual slum with their luxury homes on the face of Red Mountain as viewed by the peasants from the smog-filled valley below. Despite having ruined everybody else's view they want to keep the riff-raff out at any cost. The arrogant attitude of some of these snobs inspires unspeakable contempt. It is one of the dark sides of the sociology of contemporary Aspen. It could be that a greater sense of civic responsibility is in order. To make matters worse, newcomers named McCloskey built a big house at the natural place to enter the creek from Red Mountain and closed that access to Hunter Creek Valley. The details of the fights related to this dispute have occupied many pages of the Aspen times in the past few years.

What is one to make of a super-rich community that has made it almost impossible for the worker bees to live in Aspen, or to interfere with public use of federal land that belongs to all? The Red Mountain situation is only one example of the core problem that is tearing apart the social fabric of this lovely place.

Access 3: Smuggler Mountain. There is a jeep road beginning at the dumps for the old mines. You need to get to South Avenue from Gibson, reached from the corner where Main turns into Original. (You can also reach South by heading north on Mill Street and taking the right turn on Gibson at the T junction) Take Neal St. and turn left on Gibson, then right on South. driving past the trailer park. There you will reach the old dumps (Molly Gibson Shaft).

As an aside, the Environmental Protection Agency has been on a toot about lead in the mine tailings. So don't eat the dirt. They would dig up every gram of the stuff at great cost, no doubt liberating vast quantities of poison. As you can imagine the EPA is much hated in Aspen as an unwelcome bully and a symbol of what is wrong with our well-meaning environmental laws. Their latest burst of genius is a project to feed pigs Smuggler tailings to see if they get sick.

A sign announces Smuggler Mountain Road. Turn left. The road is deceptively smooth at first, having been tidied up in recent years. Formerly a four wheel drive road, it is now a smoothed dirt road almost accessible to ordinary cars. But wait till the first heavy rain! The road is narrow and has few turnouts. Do not look at the fantastic views if you are the driver. Passengers on the cliff side coming downhill will either be hysterical or pass out in fear.

As you can see this is my favorite way to go. After a while you stop climbing and enter a forest. Almost immediately the road branches. Straight ahead you get to the Warren Lakes. Take the left and drive a little ways through the forest. At the Forest Service boundary there is a gate, formerly closed. If so, walk down a steep hill and continue on the road for about forty minutes, where you will come to a rotting bridge and a dammed pool that was formerly the city reservoir. At this point you encounter the other (closed) road. Now walk upstream on the road, past a jungle of beaver works on the right. The stream goes off to the right as you walk upstream. Soon you will encounter a giant meadow; a few hundred yards to the right you will see tall blue spruce trees lining the stream. Personally I never fish downstream from the reservoir. There are more hikers there than I like.

Since the walk from the car is an hour, and from Aspen trailheads even longer, carrying hip boots is a problem. Trudging along in hip boots in the warm sun at the end of the day is a drag. I usually put my equipment in a back pack. Don't forget a picnic and plenty of water and beer. River water is not safe (sometimes it contains the protozoan Giardia, which can make you very sick). It does not bother me but should be mentioned. I recall having digestive problems in the past when I returned to clean water. Evidently beaver act as the host. It follows that the delicious water in Hunter creek is especially unsafe.

Hunter Valley is rich in bird life. Take binoculars.

Fishing. Now finally we get down to fishing. Hunter Creek teems with tasty brook trout usually 6-9". Their diet is mostly May flies. Somehow they are more beautiful than brook trout from other places, but that could be due to the beauty trance that is induced by this valley. In the upper reaches near Thimble Rock there are a few cutthroats that can run a little bigger. Put them back. Having crossed the meadow you will find alternating runs and pools. the latter come small and deep as well as sometimes having long tails that serve as feeding areas. Also pay close attention to undercuts and corners. As usual the larger trout will be in the front of a run emptying into a deep spot. Refer to the general instructions of Chapter 1. Remember that a short rod is better here because of the dense brush along the bank.

The dirt road continues about two miles past the reservoir. Here the stream develops a steep gradient, creating boulders and deep small pools. This stretch is harder to wade, requiring a lot of boulder hopping. Then the terrain becomes more flat for a while, with deep holes and nice runs.

Anybody who has learned to cast has only to tie on a Rio Grande King wet or an Adams to start catching trout. Dead drift upstream is best for wets; with the white wing you can track your fly and know exactly when to strike.

The population of brook trout is impressive At the end of the summer they are quite plump. One day the action was so fast that I wondered how many could be caught. In two hours I landed 80, most of them safely returned for a future occasion.

Presently the general limit is eight trout of any sort plus ten trout under eight inches. Therefore if you have a big family you can take 18 tasty trout. Since so few people have the stamina to go this far, and since there is an overpopulation problem, I don't think it hurts the river. If you wish to save some for later, or to transport them a short distance, put them in a 1/2 gallon milk container and freeze hard. Of course you will put them in a cooler with more ice for transportation. If your freezer is good you will never know that they were frozen. (p. s. Remember that there are laws governing the transportation of game fish).

In late afternoon sit on a rise to watch the western sun and wind play on the iridescent grass. If you can, plan to camp in one of the many nice places along the river. I like the spruce groves best. By crossing the river away from the trail you will have more privacy.

Finally I must warn you about the dangers posed by the giant population of Beavers along Hunter Creek. Although Hunter Creek water is now being diverted to the east slope, the beavers do a lot of diversion on their own. Many places are swampy, muddy fields with deep pits and terrible brush, stimulated to grow by who knows how many kinds of beaver water. Sometimes you will be tempted to return to the dirt road to save time. You look across several hundred feet of forest and obscure wet spots in the forest shadows. This is a field of fisherman traps set by the beavers. If you fall into this trap you will be sorry. Go upstream, or downstream, until you find a dry way out.

2.11 Castle Creek

To get to Castle Creek, go west of Aspen on Route 82 to the second stop light and turn left. the road immediately forks; the left branch goes up Castle and the right Maroon. The lower few miles of this spectacular valley are much built up, so I usually continue to the first bridge. On the way notice the Music School on the left (there are ponds with big trout milling around) and later Midnight Mine Road on the left. This eventually winds up to the top of Ajax Mountain, as the ski hill is known. You can in fact drive directly to the gondola if you find the way.

Much of the river in this long valley is a straight shot: fast and shallow, relieved by pools here and there. For orientation notice that the "first bridge" is three miles in and the "second bridge six more miles. These are convenient entry points, but there are many others. About 1 1/2 miles beyond the first bridge is Conundrum Creek, allegedly the home of cutthroats. However I have not tried it.

Three miles beyond the second bridge one finds the (barely saved) ghost town of Ashcroft. Below Ashcroft is a nice stretch of winding meadow stream just beyond Elk Mountain Lodge. You need to ask permission to fish here. In addition you should not neglect the beaver ponds. Below the first bridge there is a good one visible from the road. It has been in business for a while now after a hiatus.

From the first bridge we go downstream. Starting at the corner, where the stream turns left, there are trout (rainbows and brooks) in little pools on the east side and also in the rapids. The undercuts shaded by willows on the west side are usually good. Continuing downstream there are nice deep places. Beyond the beaver area (not visible from the river) the current rushes on relentlessly. It is dangerous to walk downstream and can be quite difficult in the jungle of brush, although a few passages exist. You will soon understand why there are few fishermen here. Finally the stream opens up, with nice pools and usually many trout. The silvery 10-12 inch rainbows are hatchery trout. They taste fine. The long time resident fish are usually shorter. After these pools there is a dam and a waterfall. Beyond it is private land.

In getting out do not be tempted to take a short cut by climbing to the road on the steep bank on the west side. I have twice got into trouble, once getting a hook in my hand and the second by not being able to complete the climb or to go back down. Sliding backwards with big dropoffs behind is no fun. In this case I got five feet from the top and waited for a cyclist to save me.

The wading in this fast deceptive current is scary in high water. In this case just go upstream from the bridge, where you can walk along the bank.

Skipping a lot of water we go to the second bridge. There are nice pools upstream. The rocks are the wrong size for easy wading, but there are fish, sometimes 12-16 inches. I have heard of bigger ones.

2.12 Maroon Creek

The remarks on Maroon Creek fishing are similar to those about Castle but the navigation is easier. There are quite a few turnoffs. If you go all the way to Maroon Lake you get the famous view of the Maroon Bells and an occasional trout. Crater Lake, one mile uphill towards the mountain, contains no fish because it is too shallow and freezes in the winter.

As a matter of convenience note that in the summer the road is closed to vehicles from 8:30 AM to 5 PM. You will need to get up early or go late. Of course you can ride the bus for $4 in 1993. The bus leaves Rubey Park terminal downtown every half hour and departs from Maroon lake on the half hour.

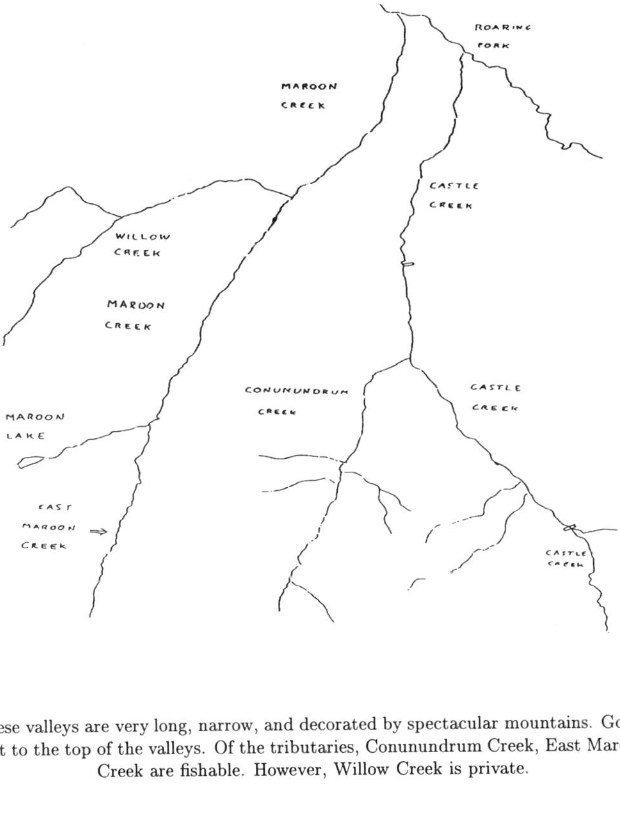

Figure 2.3: These valleys are very long, narrow, and decorated by spectacular mountains. Good paved roads take you almost to the top of the valleys. Of the tributaries, Conunundrum Creek, East Maroon and Willow Creek are fishable. However, Willow Creek is private.

2.13 Snowmass Creek

Snowmass Creek joins the Roaring Fork at Old Snowmass Village on Route 82, about nine miles downvalley from the turn to the ski resort. No doubt there are some good fish here, but there is a lot of private land. I usually drive through West Snowmass Village and take the road heading south up the ridge over to Snowmass Creek. This takes you directly to the trailhead of the pack trail. The latter goes past Snowmass Falls Ranch and then enters the Maroon Bells-Snowmass Wilderness Area. This is a wonderful place to hike and camp.

About one mile past the ranch there is a trail to the right that goes to West Maroon Creek. The trail takes you to a high meadow surrounded by mountains, very close to Capitol Peak. The stream is small but full of cutthroats. Early in the season I would not count on being able to ford Snowmass Creek.

Continuing up the trail keep your rod out since the stream is close to the trail. Beaver ponds will appear, blocking the creek and forming large pools with brook trout. As you get near Snowmass Lake be sure to fish the outlet. In the lake 20 inch cutthroats rise out of reach of a fly and usually a spinning rod too. These days you cannot camp as close to the lake as before; the shoreline was taking a beating. Turning left towards Buckskin Pass there are many waterways and ponds.

2.14 Capitol Creek

To reach this lovely valley drive west to (Old) Snowmass and head south for about three miles. Take the right fork and go straight for a few miles. On the left is a turn to the Saint Benedict Monastery. Continue about a mile and a half and take the left fork of a Y junction. If your vehicle gets you to the end you will look down into the valley to the east. Dropping down to Capitol Creek, poke around until you find the trail and a good place to cross. Early in the season the stream can be deep. Now it is an eight mile walk to the lake. There is a lot of good fishing on the way. Capitol Lake has good sized cutthroats. Watch that your backcast does not hit the rocks behind and denude your fly of its point. It is easiest to catch them when a light wind ruffles the surface. If you go all the way to the lake, you will want to camp out.

2.15 Woody Creek

Go past Upper Woody Creek Bridge coming from Aspen on Route 82 or McClain Flats Road. Go past Woody Creek Tavern (watch out for the speed control ditches) and take the first right (a very sharp turn) and head up the valley. Here there are still ranches and the scenery defines the word bucolic. There are many ranches, a few homes, and little public access. You will have to ask permission. At the end of the paved road there is a cattle guard and signs saying to stay on the road for the next 1 1/2 miles. The stream is very pretty and has cutthroats. Continuing (be careful, the road is narrow with steep cliffs on the side), the banks are unpassably steep for several miles to the collapsing town of Lenado. Basically without private access, this nice stream is therefore too difficult to fish.

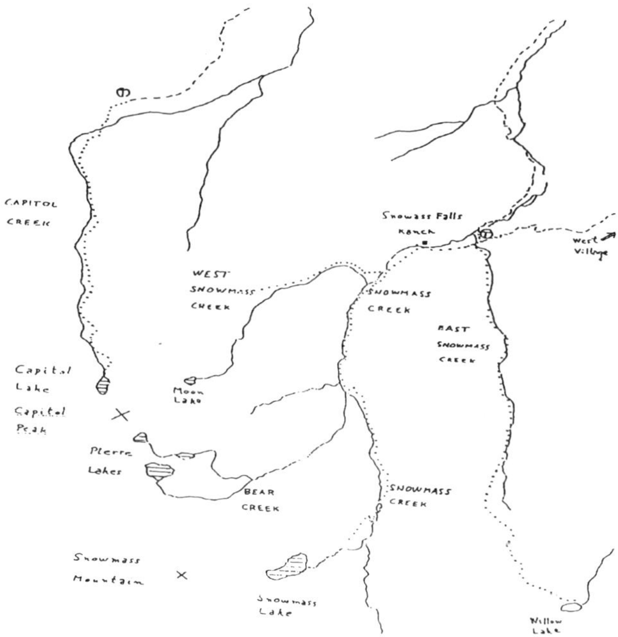

Figure 2.4: This large wilderness has lots of good fishing. Walking is the way to get in, unless you are a horse person. There are many trails, although sometimes you will have to navigate by topo. Note the trailhead to the Snowmass valley close to Snowmass Falls Ranch. Route 82 is several miles to the north.

Chapter 3

Fishing the Big Waters of the Roaring Fork

3.1 Introduction

Fishing these interesting and challenging stretches of river first of all requires information on access, and secondly the particular nature of the stretch to be approached. To deal with this we have partitioned the water into beats, most accessible in a few hours of fishing. Keep in mind that the drive-up stretches downstream from Slaughter House Bridge and upstream from Upper Woody Creek bridge receive a lot of pressure in the summer. The fish are there but become skitty after being harassed by so many fishermen. If you are willing to climb in and out of the canyon or to take a half hour hike on the old railroad bed. In some areas you will not see one other fisherman on a Sunday in the middle of August. Few rivers of this quality can offer that luxury.

As mentioned in the Introduction I have spurned the epithet "quality" in favor of "big". We will discuss the big water in fishable segments. Access is easy all the way down to the Woody Creek bridge because of the old railway bed, now the Rio Grande Trail that I have denounced already. Watch out for reckless bicycles; don't get carried away and hurt your rod by pointing it at noses or whipping the culprit. Then there are kayakers and boat people who smile foolishly as they bounce by, scaring the fish. I often had the urge to use my fantastic casting skills to catch one of these guys by the ear lobe. Hang gliders are OK if they don't fall in my part of the river.

At present all this water is "catch and release". Also there are restrictions on possession both above and below the quality section. Before you buy your license make sure that the establishment has a copy of the regulations since they have become increasingly complicated and restrictive. Although having never been asked for my license in thirty years of fishing, I suspect that the day you have strayed will be the day you will be caught.

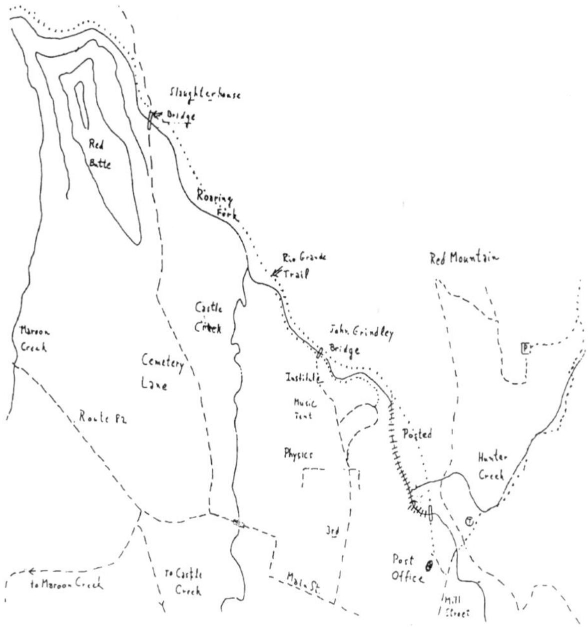

Figure 3.1: On this map we see three main tributaries of the main Fork. In addition access to Hunter Creek by foot is shown. Except for the posted stretch in town downstream from the post office, the river is very easy to reach. Note the road to the north of Slaughter House Bridge. It is a convenient way to go to the Upper Woody Creek bridge.

3.2 From Hallam Lake to Slaughterhouse Bridge

Behind the post office off Mill Street (downhill from the Jerome Hotel on the left) you will find the entrance to the Roaring Fork Trail. (To the left is Hallam lake, now a nature preserve). Here the trail is paved and wide. Proceeding to the river, you are at the upstream beginning of the quality water. Soon Hunter Creek joins the Fork. Looking downstream, you will feel some adrenalin as you see signs saying "keep out". This restriction continues down to the Roaring Fork Road entrance that I will discuss.

Now we come to a lovely little secret: some of the most pleasant and uncrowded water on the Fork is right inside the city limits. Go north on Third until you come to Gillespie. The music tent is visible; bear to the right on Roaring Fork Road and ignore the summer girls guarding the Aspen Institute parking lot. Down Gillespie to the left is a complex of buildings called the Aspen Center for Physics, whose practical organization has been taken care of by Sally Mencimer for more years than she would like for me to reveal. In the summer about ninety abstracted scientists speculate here about the mysteries of the universe. Some of these have had the sense to take up trout fishing, which is a better way to approach problems of the universe.

As you proceed past vast houses, empty much of the year, you will come to a corner where you turn left. Notice the gate and the fence separating the dirt road from the County Open Space that goes along the river. There are couple of convenient pulloffs along here. Go through the gate and you will find a lovely trail soon going downstream about forty feet above the river. Sometimes you can see giant trout that will have disappeared by the time you get there. Soon the trail descends through a damp forest. Look for raspberries in August. After a little over five minutes (unless you stopped for berries) notice a little rivulet and evidence of beaver work. Soon you will see a little trail to the river; recently the banks have become overgrown. You should come to a long, mostly shallow pool with moderate current; this is a good place to cross. Upstream you should see a big rock on the right with a deep run on the north side. I once nabbed a 19" brown there (deep with a hare's ear) and have lost bigger ones. Anyway, first fish the mentioned pools, then cross and fish upstream to the forbidden territory. You will find many interesting places to fish, especially from the north side. There are many small pocket pools which are hard to fish because of the need to cast over a complex current pattern at distance to avoid alarming the trout. At night you can have the joy of watching a cunning trout wrap your leader around your leg to break off.

Back to the entry point, look south and you will see a new bridge; the John Grindley Bridge. I consider this bridge architecturally elegant, an opinion you can ignore. This connects the nice trail in the forest on the south side to the Rio Grande Trail which goes on to and beyond Slaughterhouse Bridge. If you go uphill a steep trail takes you to the parking lot near the Institute, a short distance from your car.

Mr. Grindley is survived by his wife Betty Grindley, a nice lady who helps in many local cultural activities. Her house, on Pitkin Green Road, overlooks the river. Her son Josh, is an astrophysicist at Harvard. He often comes to the Physics Center. I once bought a Czech violin from Mrs. Grindley after a party in an apartment downstairs. The next morning I was relieved that the party had not compromised my musical judgement.

Going downstream from the bridge you will find wadeable water that is very pleasant. Soon Castle Creek joins the fork; this is a good place to look for big trout.

The entire distance is about one mile. If you have no a car, you can take the bus (Snowbunny, presently free) to Red Butte Drive on Cemetery Lane. It is a short walk downhill to the Roaring Fork Trail, fish up to Roaring Fork Road, walk to the Music Tent bus on Gillespie in front of the tent, and go home rather tired. Save most of the day for the full treatment, since you will dawdle. Because of the easy access you can have instead a two hour session if you wish.

This stretch is excellent for brown trout. Besides boulder guarding, they often are in shallow water near the bank, especially at the head of tiny pools where a little stream of water comes in. They are usually thinking of food even when the rainbows are asleep.

Flies In this area more variety can be necessary than in the heavier waters downstream

Size 14-18 dark caddis, pink ladies, light Cahill, mosquito, irresistible, quill Gordon, red quill, blue winged olive, are examples that I have found useful. Usually you will not need them.

- **Interlude: Musical Trout**

  George Stranahan, an Aspen notable who lured me to Aspen in 1962 when he established the Physics Center, is an interesting character about whom more will be said later. I had been directed by him to various areas, one of which was the tent water. Here I found perfect places. My favorite is a long pool (described above), a waterfall feeding into a deep run sliding past a long rock, two side entries and a long tail. At the head are two vertical rocks where trout can rest, waiting to ambush prey coming over the falls. My first cast to the vertical rock produced a savage strike from an enormous brown, which broke my leader. However, next year I got a 19" brown in the deep run. Of course the fly was a gold-ribbed hare's ear nymph.

  After the defective sewage plant in town was moved downriver the musical fish were not so common or large. Nobody wants to hear my theory that a little nitrogen drip fertilized the river. This theory is just as good as newspaper medical advice.

  The Fork is basically a rainbow river despite the above stories. The browns typically come upstream from the Colorado River in the fall but there are always a few in residence. Also some brook trout and an occasional cutthroat, usually hybrid. And a few whitefish.

  The biggest rainbow I ever saw from the Fork was caught by a tenor in the summer opera of the MAA (Music Associates of Aspen). As I came home I heard loud cries of joy next door and went to see what was going on. Here a six pound beautifully colored female rainbow flopped in the kitchen sink. I wonder what he was singing. He said that he heard drizzling rain and went the river armed with a very large black ant. The location was in what I call "pipe rock pool" (see next section). I can figure out the year because I was living at 2nd and Francis. This brings us to another stretch of the Fork that I like.

3.3 Slaughterhouse Bridge to Maroon Creek

Yes, there was once a slaughterhouse here. To reach the bridge, go west of town on Main Street and turn right on Cemetery Lane. Watch out for speed traps. Across the Fork on the left is a parking lot. The half mile downstream is perhaps the most fished part of the river. Wait until September when things quiet down. Walking downstream, you will see Red Butte on your left. It has some interesting geological history which I forget. Notice the slanted rocks going down to and into the river. This provides a very interesting corrugated bottom in many spots (the ridges mostly perpendicular to the stream flow). I believe that this must provide good resting places for the trout. All along this part of the river we find rough water with falls and deep holes, as well as long deep pools.

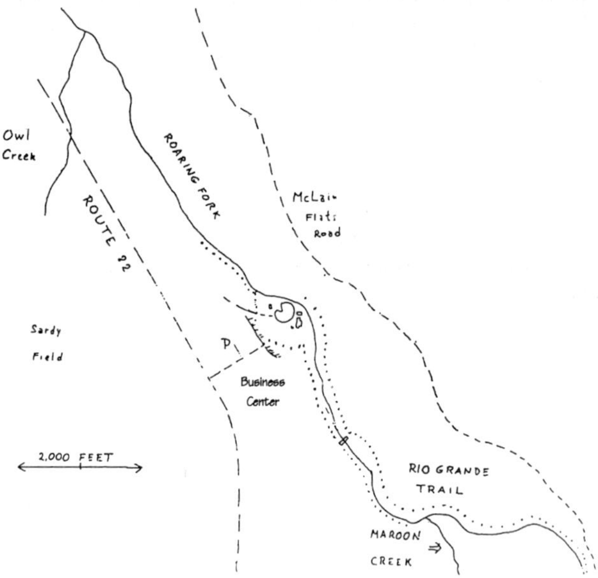

Figure 3.2: This map is self-explanatory. Note the bridge midway between Maroon Creek and the business center. The ponds are settling ponds for the sewage plant. Although a gravel road goes there your car is not welcome. The same is true for the next road west, which leads to a large bus center for the valley.

The peace and safety of the trail are compromised by hordes of joggers, dogs and bicycles, not to mention some horses. In the river the sight of so many rafts and kayaks inspire the ordinary peaceful angler to thoughts of revenge.

After ten minutes walking downstream you will come to a rock overhang. Look at the rock formations to your right. Access from the south in this section is physically impossible. As you keep going, you will find deep holes, which seem to have very experienced trout as inhabitants. Keep going and in five minutes or so you will come to a fairly big waterfall near some slate cliffs. Often there are trout cruising in plain sight. The boil at the foot of the falls is often good. Ignore Maroon Creek and fish back upstream. Notice the pond on the north side, which used to have 10" brook trout. Here you will be attacked by swarms of Tiger Swallowtails.

Pay special attention to boiling pockets behind medium sized rocks in a fast current. Often a big trout is lurking within. Drop a nymph in (add a touch of lead) and let it swim a bit. In early season when the water is murky and cold I have found that a large Black Wulff (pink body and black wings) does the trick after being sucked under.

Skipping the chance to write about a lot of nice water, I return to the overhang. Just a little way upstream is one of my favorite places in the entire river, although it has suffered from too many fisherman hovering, and careening down the steep bank. One boulder has a short pipe stuck in it. I don't know why. Anyway I call this pool "pipe-rock pool". This is where the tenor caught the six-pounder mentioned in "Musical Trout". The bottom of this pool is rugged; there are many deep runs harboring many large trout. On a bright day you can see big trout nymphing a few feet under. This calls for a hare's ear or muskrat nymph with enough lead to get deep. At least by evening there should be many rises. To fish this pool go downstream to the foot of the pool, slide down and wade upstream carefully in the edge of the stream. Gradually work up to the deep runs and the shallower slower water on the opposite side. Now you can cast your nymph (with lead on the leader) dead-drift straight upstream into the deep run, or diagonally across. Hopefully some fool will not spook the fish by walking straight up to the edge. Be aware that the boulders on your side are undercut, and that survivor trout wiggle under them sideways to get rid of the hook. Do anything possible to keep them from doing it. Here you can't move and the water is somewhat fast. Good luck. Going upstream there are nice places that are hard to describe. Often the little spots are better than bigger ones. These you have to learn by experience.

Pause. Meal Hours in the Fork.

Naturally there are many variations in the feeding behavior of the trout. On a typical July-August day, my experience is as follows. Until 12:30, the water is usually cold. Then the insect activity, if any, begins and it is lunch time. About three it may be all over and the fish take a nap until around six, unless a good hatch continues. Although there are some tricks to arouse the slumbering trout, it can be very frustrating. Also note that no visible hatch does not mean that the trout are not feeding on nymphs near the bottom. This is why it is essential to learn the weighted leader method of nymphing.

The evening rise is problematic. Sometimes by 6:30 it is clear that there will a lot of action. Other nights the river is silent as a tomb. When a good evening hatch occurs don't waste time. You can catch a large trout every few minutes. If it continues into the darkness, stay. Make sure that you have a flashlight and know the way home. Usually you will be the only crazy fisherman out so late. At a certain time I cut off my gentleman leader and go to 1X with a size six Royal Wulff. No messing around with minnows now as the giants begin to roam. This sort of fishing induces real mania as the trout plop all around, while you strike repeatedly at nothing. Then a big swirl and your fly is gone, along with 1X leader.

3.4 From Maroon Creek to Airport Business Center

The Fork is now large, fast and deep. First I will describe separately the fishing from the south and the north banks of this stretch, which is less than a mile long. Since there is now a bridge at midpoint, you can easily cross from either side to the other. However it would be too confusing to list all the options.

To begin I describe my favorite beat. Drive west on route 82 almost to the airport, turning right into the Airport Business Center at the Chevron Station. If you go straight you will be at the trailhead at the edge of the canyon but will not be able to park. The next (dirt) road back goes west to a parking lot by an athletic field. A five minute walk takes you to the canyon edge, overlooking the sewage disposal plant. There is a steep dirt trail down the hill (about 200 feet of elevation change) which skirts the plant on the right and deposits you on an dirt track which parallels the river all the way up to Maroon Creek. Across the river there is a high shale cliff on the side of which is the Rio Grande Trail, now rather high above the river. Still higher is McClain Flats Road, which goes to Woody Creek.

3.4.1 Great Pool #1

Straight ahead there is a very long and wide riffle, with nice deeper runs around boulders. Although these are interesting, they are secondary to stalking the giants in the pools upstream from here. Continue upstream a few hundred feet and you will discern a long deep pool with interesting currents and rock structures. This can be one of the best pools in the river, although fishing pressure has increased due to easy access and word of mouth. Identify the foot of the pool and sneak in, continuing slowly upstream crouching as you go. Because of the dense spruce stand along the bank you are restricted to backhand and roll casts. Soon you come to a little sand bar. Get on your hands and knees to simultaneously pray to the fish god Ichthyus and to flip a fly into the nearby currents. The fish may be only six feet from you but can be fooled. Next you are entitled to get up on your knees and cast farther. The brush behind you is not so bad here if you are careful. After this you can stand up, partially obscured by some boulders along the bank. Usually I begin with a dry even if there is no surface activity. Often the trick is a nymph (caddis or gold ribbed hare's ear) with weighted leader. Cast successively farther into the various currents.

The fast deep run in the center breaks up into many components. At the head of the pool on the right is a nice whirlpool. On a bright day you can see several trout here, usually nymphing. There is a big boulder at the edge of the incoming current. Monsters live there, but usually come out only at dark.

Poem: Trout Rivers I

Every spring
Trout run upstream in my veins

Fish are beautiful primordial beasts

My hunter instinct leads me to kill
And to eat

As in catching women
You float a false image over them
Once inspected
It is taken or rejected
Yet with women
The catcher is caught

While dwelling on Great Pool #1 we describe some lairs that are best reached from the north side. Just yesterday (September 4) I checked that it is an easy wade now from one side to the other) at the foot of this pool. There is usually some action in the runs in the foot of the pool. In midpool the current spreads out, with a long run where large trout rise everywhere during a good hatch. It is barely possible to reach from the south side, but you have to cope with drag. Having exhausted this, wade in as far as possible and fish deep in the run. At the head there is a rocky bank with a deep dropoff where nice fish often lurk.

This fine pool is a complete fishing lesson in itself. You might suddenly notice that you have spent an hour.

Remember that sometimes the trout have left the pool to go foraging in the more shallow waters above and below.

Continue to fish upstream on the south side.. Remember that the nature of the fishing changes when the water goes down as the season progresses. You can reach spots previously too difficult because of the currents, and I believe that the trout also rearrange their living accommodations as the season water conditions change.

Continue upstream. Remember that the fish, sometimes large ones, often have taken up residence in small pockets and otherwise unlikely places. Through experience I have found many of them, and must admit that the superficial appearance of fishiness is not reliable. In such water fish quickly but cover as much of the river as you can.

3.4.2 Great Pool # 2

In another 100 yards you will come to another outstanding complex of boulders and deep runs. Start at the foot of the pool casting upstream and also across. Peek over the first rock upstream to your right and drop your fly just in front. Surely there is a trout there with your name on it. Sink a nymph deep in the current by the long boulder about 20 feet in front of you. Continue up, not missing any little pocket. On the far side you will notice a straight section of rocky bank. In the evening large trout lurk along this bank. This is the scene of the "trout with the mouse" incident described earlier. Upstream is a big and deep whirlpool. These spots must be fished from the north side.

Continuing, there are many good places to fish. In addition, this area is dipper heaven. These fascinating birds enjoy doing pushups on the rocks, only to jump in the water and reappear with a beak full of bugs. A real treat is to see a flotilla of babies bobbing in the current. I wonder how many end up in the stomachs of trout. Along here under the overhanging spruce there is a "soft rock" where I have a beer or take a nap As Carlos Castaneda points out, it is important for your inner strength to identify your "spots". This is one of mine.

Soon you come to a fine wooden bridge. Here you can cross if you want. But continue to Great Pool #3. If you are on the road you will soon come to a gate which in former times had a no trespassing sign. Here in any case, I go back to the river and wade carefully upstream, fishing all the runs and pockets. In the near distance you will see a broad river with a substantial round boulder a little off center to the left. This was the home of the late Moby Trout, King of the Pool. His survivors are many and large.

At feeding time you might see a dozen sizeable trout feeding. Most commonly they are catching nymphs from six inches to six feet in depth. Perhaps it is still deeper, but I can't see any farther. Except late in the emergence, dry flies are not very good, since the trout are looking down. The gold ribbed hare's ear is reliable. Still better during and after the main hatch in late July is the so-called "green drake". To imitate this Ephemerella I have designed a variant of Schwiebert's "dark great red quill". I call this fly the Sir Francis Drake, after the great mariner (see Section 5.4). This fly, tied on a lightweight size ten 1X long hook, works wonders. It can be fished wet, which is often the best strategy. Even drag does not deter the fish. The fly brings big trout off the bottom long after the main hatch has come and gone. I always try to get to Aspen by July 31 to catch this hatch at its peak.

According to our earlier plan you are wading up to the pool a little to the right of the middle. At the foot start fishing, gradually throwing long weighted lines into the current. You cannot go far because of the depth. Fish the tails and try to fish the side of all boulders. Now wade in the shallows to the right and cast over the current. This is difficult. Now climb to the top of the next big boulder, stand up (!) and cast in and over the current to the feeding fish. Long casts and the rough current will keep you from being discovered for a while. If you get a big one on slide down the rock back into the river and hurry downstream and escort the fish to sidewaters and then into your net. Otherwise forget it.

In past years I often fished this pool in the evening with Rubby Sherr, formerly a professor at Princeton. After being broken off a number of times we began to call this trout by the name "Moby Trout". We never saw, only felt, this monster.

Upstream there are several more good spots. Finally before Maroon Creek there is a very long run where many trout hang out. Usually they are not especially big, but there are many.

Having lavished so much time describing the south side, I will not give so much coverage to the north side.

If you are coming from Slaughterhouse bridge, it is less than a half hour walk on the old railroad bed to Great Pool #1. It can be very hot in the summer so take lots of liquid. If you wear waders roll them down a bit. The easiest access is from the Business Center, using the bridge. You will know that it is time to go down into the canyon when you see the sewage treatment plant a little to the right, and a very long rapids further to the right. There is also a little water fall coming down from the cliff above. The canyon sides are shale chips here; it is easy to side slip down, perhaps grabbing at little oak trees for reassurance. You should now be at Great Pool #1. Now fish upstream through some truly beautiful water. At one place, as part of a very complicated pool structure, there is a giant whirr. On the left there is a rocky bank with a shallow run harboring big fish in the evening, with or without mouse.

The Royal Wulff also brings the fish alive on those afternoons when curfew seems to be in effect. Sizes 10-14 on a 12 foot 4X leader. (The air resistance of these flies requires a heavier leader). Or a large wet Rio Grande King retrieved from downstream can be very effective at times,in sizes 10-14.

- **Interlude: The Trout With the Mouse**

  One night near dusk the fish went on a feeding frenzy. In half hour I landed four fat rainbows around sixteen inches. Then I noticed a rise near the bank upstream. After an enormous battle, during which the fish jumped out my net twice, I pounced on a heavy 20" brown. Having dispatched the brown, I performed an adrenalin-drenched dance around its body under the full moon. Luckily nobody saw this primitive display. As I meandered along the railroad tracks toward Aspen I encountered two fishermen who inquired about my luck. I pulled out the brown and let them peek at the others. "How did you do it ?" Modestly I explained that you need a large (say size six) Royal Wulff pattern and a very strong leader. The Royal Wulff looks like no insect I ever saw; rather it looks like something out the English Court. The white wings make it easier to see them at night.

  To finish the story of brown trout, I opened its stomach to see what it had been eating. Out popped a minnow, no surprise. Then I noticed some red meat. Squeezing both ends of the stomach I observed two halves of a mouse. The mouse must made a last moonlight swim from a nest in the rubble lining the bank. For punishment we stuffed the brown with wild rice, wrapped him in foil and baked him. Quite a delicacy We did leave out the mouse, however.

Poem: Trout Rivers II

Twenty Thousand trout in my bag
Over the Aspen years

I know how my life will end

By falling in the river
Pulled by Moby Trout into the torrent

Then ten thousand trout
will pull off my flesh
My bones will scatter on the rocks

Serves me right

3.5 Business Center to Shale Bluffs

The Fork bends around the sewage ponds. In the corner there is a medium sized pool where there is always some action. Throw a dry fly against the rocks in hope of getting a brown. You will need to make a loose cast because of the swift current in the middle of the pool. If you let your line straighten out in the tail a brookie is likely snap off your fly. Notice the steep high shale slopes above. It is the devil to climb. Proceeding downstream several hundred yards in the shallow flat riffle notice the undercut bank on the far side. At the next turn to the left there are deep holes and big fish.

Usually I fish the south side of this stretch. It is easier to reach. Ordinarily I do not reach this section from the upstream side of the treatment plant. Instead, from the parking lot at the athletic field I walk directly to the edge of the canyon. Looking down, you will see the treatment plant. Stay to the left and pick your way down the slope, which is populated sparsely with scrub oak. Watch out for the barbed wire. You can walk around it to the left. Now on a gravel road, you can find a marked fisherman's trail down to the river. This comes out just at the tail of the big pool described below. That description imagined that we were coming from upstream skirting the south side of the river.

(Another option is to walk along the top downstream until the bus depot, and go down in order to fish upstream. This can be tough going. The lack of a trail, the many barriers, and too much rock hopping can make this a strenuous trip. Now return to our description of the downstream walk.)

Having reached the river turn right and find a very long deep pool. This was another great pool before the effluent (affluent?) came. Perhaps the oxygen level is low here. If you go past the head of the pool on the right you will find four small pipes releasing sewage plant effluent.into the river. This dumps into a tiny pool where in former times you could get several trout 12-18 inches without moving. Returning to the big pool fish it with deep nymphs (sometimes there is a hatch). Whitefish have invaded the giant pool, although some trout remain. However, within a few hundred yards you again come to good trout fishing. Don't ignore the riffles.

Below one you will find a small deep pool with a spherical boulder in the middle. Over the years this has been a reliable source of action Continuing downstream, fishing through some nice runs you will come to a pool with a giant boulder on the left and deep water in the front. This qualifies as a great pool but is hard to fish. This pool holds very large trout. Weight your nymph and pray. The casting is difficult, especially casting upstream or across with a wall of spruce and willows along the bank. Sometimes there will be rising trout on the far side. This requires long casts but can be managed. Continuing downstream you will encounter such a fascinating and bewildering.array of water types that I could not describe them well. Believe me, there are trout here. Some days you will not believe it.

In this stretch of water caddis and gold ribbed hare's ear nymphs are very good. In addition a large (10-12) wet Rio Grande King is very effective. This is especially true when the water is cold or when your usual flies don't work. In addition to usual (weighted) nymph methods, you can retrieve it against the current. Later you can tell your friends that it was a size 28 double-breasted mattress-thrasher instead of the humble RGK.

When you finish fishing you might opt to climb up the steep canyon wall instead of retracing. If you do, pick a good location and be prepared to take your time. The slope is fairly dense with scrub oak, though nothing like the nightmare jungle on the upstream side of the Business Center. If you have an accident on the slope, which is unlikely, you will be hard to find unless you take your trusty whistle. In any case don't try this in the dark.

Finally, take plenty of fluids of your choice. You will dry out fast in the low humidity, sometimes hot weather, and perhaps overexertion.

- **Interlude: Sneak Attack by a Park Bear**

  I have encountered no bears around Aspen, so we will have to go elsewhere for this story. It was 1965. My family (with wife and three kids) were camped in Yellowstone near the most remote loop in the campground. This was real luxury, with our 10' by 18' Sears tent, a table and a grill. Every afternoon new campers would arrive and set up shop. Unfortunately for them this was the territory of a smart bear. He marched in, scared off the newcomers, who left in terror or sat shivering in their cars all night. This smart bear knew how to open all coolers by lifting them up repeatedly and dropping them on the table.

  I was grateful to the bear for keeping our spot free of competition.

  After about a week our peaceful coexistence came to an end. I had gone over to the Yellowstone river just above the falls. The river is very broad at this point. There must have been a hundred bait fishermen lined up along the bank. The river was filled with many insects and constantly rising trout. Wading in with my trusty 7' Shakespeare Wonder-rod, I marveled at the long casts I could make, despite the wall of trees behind. However after an hour no takers were found. Scooping up fly after fly I could not figure out what they were taking. I was ready to quit when I noticed a large ephemerella coast by. Then another. Aha! Just the fly whose imitation I had devised for the Fork hatch.

  Next came a cutthroat coming upstream going slurp, slurp, slurp, methodically turning to the left and then the right. Casting a few feet ahead he slurped in my fly. Only 13". Then another came, with the same result. It did not fight hard, so when this two foot massive trout almost fell out of my net I decided it was time to quit. Naturally a crowd of salmon-eggers gathered around to find the secret. I modestly suggested that they needed to take up fly fishing and tie their own flies.

  Anyway, back at the tent I built a big fire and threw my trophy on the grill After a while I heard a noise and looked behind to see my friend the bear about ten feet away, hoping to get my trout. With instinct dominating sense I jumped up and chased the bear away, using the classic technique of banging two skillets together. The bear did not return.

  Moral #1 Trout can be very picky.

  Moral #2 Always take two skillets camping. Iron is best.

  I almost forgot to tell about the theft of a trout by a dog. The scene of this crime was Salmon Creek, an inlet to Cayuga lake near Ithaca, New York. Every spring, the rainbows run upstream to spawn. April 1 is the first day of the season, with the banks lined with fisherman desperate for some action. The biggest danger to these sex-crazed trout is getting stepped on. One day I had a small one, about 15", enough for dinner. I put it down.to take off my gear. Looking back I saw a dog flip the trout in the air, deftly catch it crosswise and run to a nearby trailer, where his mistress opened the door to collect the purloined trout. I learned to look out for this dog, whose job was clearly to steal trout from the unwary.

Moral #3 Never trust a dog. It will run from a bear or steal your trout.

3.6 Shale Bluffs to Upper Woody Creek Bridge

There is parking at the Upper Woody Creek Bridge, especially on the north side in the Wilton Jaffee, Sr. Park on the north side. This strip of land goes a ways along the river upstream from the bridge. If you go east on McClain Flats road you will see Mr. Jaffee's ranch on the left hand side of the road. This road comes straight from Slaughterhouse Bridge and is a prettier and safer way to get to the Upper Woody Creek Bridge. Be sure to have lunch at George Stranahan's famous Woody Creek Tavern, a short distance into the metropolis of Woody Creek. George's ranch, the Flying Dog Ranch, is a little bit up the main road along Woody Creek. In downtown Aspen George runs the Flying Dog Pub, where he makes his own beer. He recently won a national prize for his brand "Doggie Style". Woody Creek houses a number of characters, including Hunter Thompson, but George can hold his own. . The bridge signals the end of the big water, where presently everything has to be thrown back. Upstream of the bridge the river is about 60 feet wide, with long deep runs alternating with pools and pocket water. Although it is possible to go upstream on the south bank, which is less fished, most fishermen follow the path on the north side. It is a fifteen minute walk from parking to the water directly under the shale bluffs, high above. From the appearance of the path you will correctly surmise that thousands of fishermen use it every summer. I recommend walking to the shale bluffs before beginning to fish upstream.

The following story is an example of what too much adrenalin can do to you. One day I spotted a very deep pocket with a boulder in the middle. Perfect. After no response I put on my trout detector, a very old trick. This is a dragonfly imitation with a 4" wingspread. Instantly a giant brown came up to inspect this ridiculous creature. The next day I snuck in my gold-ribbed hare's ear and King Brown rocketed off downstream with it. Desperate, I jumped into the raging torrent with him. He dived under a boulder which I embraced not knowing that I was in trouble. Feeling under the boulder I tried to get a grip on this big trout. Of course he wore my leader off and escaped.

3.7 Safety in the Big River

Access can be very difficult from the south downstream from Slaughterhouse Bridge, except along the Rio Grande trail and the old railroad bed. On the north side the old railroad bed goes all the way from Aspen to Upper Woody Creek Bridge. Even so, it can be difficult to go down (and up) to and from the Fork due to steepness and brush.

During snowmelt, which may go on until mid August, it is difficult to cross in boots or waders from one side to the other. You can fall in the river and find your waders full of water and unmanageable. Or, your hip boots will fill with icy water. You can stabilize better by using a wading staff to give you three legs. Often you can simply appropriate a strong dead branch along the stream. Watch out for splinters.

Sunburn is likely if you are not already tanned. Some years, and always in some swampy areas, there are many mosquitoes. Be sure to take the proper ointments.

Once I scratched my cornea on a branch while sliding downhill on my behind. Two days of pain on the couch. The scrub oak and occasional dropoffs are also a nuisance, One moonless night I scrambled out of the valley in a dense thicket of oaks. Barely made it. No trails downstream from the Business Center means a lot of boulder hopping and wading. The rocks can be slippery no matter what space age boot you have. Practice falling.

Do not drink the water. It is all right to cool your beer in it, though, as countless experiments have verified. Put a heavy rock securely over your beer or it will end up in Glenwood Springs.

Another hazard that you might not anticipate is ticks lurking in the brush, hoping for a taste of fisherman's blood. I found this to be mostly a June problem, and localized to the lower elevations, where livestock are common. You will have to figure out for yourself how to get them out.

There is a debate about the merits of hip boots vs. waders. Many contemporary waders are like girdles. In hot weather they are anathema. They feel like wet suits but the wet is inside. (Luckily I have a "pair" that are thin, lightweight and loose). It is true that you can sometimes get a better angle on a cast with waders. I usually use hip boots. which are cooler (especially after spilling a little water inside) and easier to maneuver. This is also a good way to soften up old feet. The practical thing is to use hip boots for small streams and waders for the deeper water of the Fork. I have also seen a compromise, which covers you up to the waist. With any of these you can get into fast currents that are a real hazard.

Once in a while we all get impaled by a hook. My first experience with this was as a boy in Indiana, when my father accidentally lodged a Hawaiian Wiggler #3 in the top of my skull. After nearly overturning the boat he took me to the doctor, ruining a fishing trip and leading to an unwanted tetanus shot. By contrast the trout hooks are like mosquitoes. Usually it is no problem, just snipping the hook and pulling backwards. You will learn that some places don't hurt. Save the removal until you see some friends. Pull it out with a pair of pliers as they faint and admire your courage. Sometimes some bourbon taken internally before the removal will ward off infection. Sure better than a tetanus shot, and safer than falling in love.

- **Interlude: Trout Fishing And Affirmative Action**

  In these politically correct times it is difficult to explain that men go fishing to be alone, and to meditate in peaceful places. Trout fishing is still a male sport, and will probably stay that way. Even when heading out with friends, it is best to split up and meet later.

  Although there are some women interested in fishing it is still unusual, although I have found that they are quicker to master the rhythm of flycasting than men. And they are really good at tying flies, as a picture of Dan Bailey's Harem will prove.

The handwriting is on the wall. The Cortland firm has just issued a book especially for women trout fishermen. Local guides now offer woman-only classes.

Chapter 4

Tips for Fishing the Big Water

4.1 Introduction

This section should be entitled "How to Fish Good". a Rabelaisian extravaganza by the noted author Milford "Stanley" Poltroon. This riotous, slightly crude work will inspire shrieks of laughter in real fishermen, while others look on with bewilderment. It is a good way to detect that distressing subspecies, the yuppie fisherman, who takes himself very seriously, rarely catches fish and doesn't drink whiskey. This new breed jangles as it lumbers down the river encumbered by a full set of Orvis gadgets. Don't get me wrong, I have several Orvis rods and reels. (The noise reminds me of Spanish goat bells. Imagine the trout smirking as they say: here comes another one). They sleep with the latest copy of Trout Unlimited under the pillow (don't get me wrong, I was once president of a chapter). They rush out to buy or tie the new puffer fly whose only role is to keep fishing magazines alive.

Meanwhile I am dreaming of a skillet full of little brook trout, wild mushrooms simmering in butter, and raspberries that I picked myself. Unfortunately the days of Toklat sourdough, a symbol Aspen in the days of Stuart Mace, seem to be gone. For years friends stored the precious starter in the fridge and shared it when somebody ran out. In any case the animal satisfaction is hard to beat.

Although the Fork is predominantly a caddis river there are important may fly hatches. In the early season there are giant stoneflies that attract a lot of attention. Most of the time the fish are asleep, so it seems, but their feeding habits focus on nymphs, either deep or actually trying to emerge. Finally when the hatch is really going they often but not always rise profusely. In side-water pockets there are usually browns or small brook trout ready to strike at any time.

Although the trout might have an early breakfast that I would never see, there is usually not much action until noon, when the water warms, awakening the insects. About three a siesta occurs, and ordinarily you might as well take a nap too, except for special occasions. By six they might start feeding again, or they might not. In any case this might be surface action.

About 6:30 some evening hatches begin and larger trout move to the surface. As the light dims, switch to bigger flies and heavier leaders. Carry a flashlight. Some evenings the hatch intensifies as darkness arrives; at other times the action stops altogether.

Suppose that you hook a really big trout. This trout usually has some experience in escaping, or natural selection took care of the ones with poor defenses. The first thing this fish will do is set off on a long run downstream in a fast current. To restrain the trout is usually fatal; it will turn its broad side to the current and wiggle the hook out, or just break off. Sometimes big hooks and strong leaders are the answer, but that only works in darkness or almost darkness.

The only solution I know is to run down even with or downstream from the fish, and maneuver it into quieter water. I am slightly ashamed that sometimes I scoop up a fish thus surprised, long before its fighting strength is gone. This compensates for the fact that often I never even get to see these monster trout before they get away.

The need to run is a reason to keep your surplus line supply neat. As you know, you should wind in all your surplus line to fight a serious trout. Work him right off the reel with the right drag set to allow sudden runs.

By the way, it helps to be young in order to run properly over the boulders. Old age and treachery do not always overcome youth and talent.

A second leading cause of big trout loss is their quick ability to dive under rocks, especially those with lots of branches jammed underneath. I find it basically impossible to solve this problem.

4.2 Mountain Weather

The sudden changes that occur in mountain weather are a constant source of surprise. And then there are days with calm gray drizzle. On totally overcast days the water from the tributaries will remain cold, and so will the main river. Fish deep with nymphs.

Storms come in at least two kinds. In the summer, thunderstorms well-equipped with lightning bolts provide an opportunity to hide under a spruce to enjoy a beer. But don't pick the tallest spruce, the lightning is real and dangerous. I love these storms.

Sometimes there will be a flurry of activity before the storm strikes. Fronts, however are another matter. Sometimes there will be wild feeding until a big front hits. The fish seem to know that the water will be dirty for several days. However, fronts arriving from the east are deadly. With practice, one can predict the arrival of an eastern front by sudden cessation of fishing.

Fishing in the rain, especially a light one, is recommended unless the wind is up. Large fish that would recognize the enemy cannot when the surface is rumpled.

Finally be prepared for the sudden appearance of muddy water due to cloudbursts in mountain tributary valleys. Iron mines at the top of Castle creek can pollute the Fork as well as occasional construction. Large woolly worms, green or black, are my favorites in these circumstances.

Then there are warm days with no clouds, when the bright light illuminates the stream bottom. Every hatching insect is visible and so are you.

In September the Aspen trees become orange yellow, the scrub oak red and the cottonwoods a yellow of a different kind. Early snow paints the mountains white. The sharp cool air smells of fall and signals the end of another year. The streams are low, colder every night, waiting for the cycle to begin again.

Poem: Mountain Weather

Suddenly
The mountain sky darkens
Trees bend in the wind
Birds hide

Rumbles of distant thunder
Predict lightning strikes
Changing the smell of the air

Suddenly a torrent of rain
Hits the high valley

Lightning tears the sky apart
I hide under a tree

Soon flowers will bloom
Soon mushrooms will come

4.3 Equipment

Consult Section 1.3 on equipment useful for small streams. Here we add a few remarks pertinent to the different or additional things needed to fish the big water.

Because of the rapid, swirling currents of the Fork, the trout seem to lie deeper or closer to the many rocks that divert the current than in a typical trout stream, if there is such a thing. The ability to cast fairly long distances with leaded line is essential. There are some excellent photos and text in Odier's book on this subject. In addition see Fothergill's chapter in "Masters on the Nymph". I like a nine foot, 4 weight graphite rod, equipped with a double tapered line. This rod should be strong but not too fast, in order to accomodate the special needs of casting with weight. If you use a standard action rod you might want a six-weight rig. I do not like sinking lines in a fast river. For leader, I use a knotted Dan Bailey "Platlon" tapered leader at least 12 feet long. Tippet size should be no heavier than 5X unless it is dark or the weather is windy or the water murky. I never bother with strike detectors; keep it simple.

Beware of knocking the graphite rod on a rock; it is might shatter later just when you need it. There are many good reels but some cannot withstand your falling on them. The reel should allow sudden runs without jamming. In a moderately deep pool cast upstream to likely spots. Fortunately quality graphite rods do a very nice job with dry flies. The length of the rod helps stretch over complicated currents. In addition the length of the rod makes long roll casts easy.

Landing nets are a necessary curse of our sport. When the net is connected to your neck by a string on a spring, your back turns black and blue as the net catches in the brush and then lets go. Various gadgets are appearing to alleviate this problem. Your choice.

A thermometer helps you to make little graphs at the end of the season. Perhaps you can learn more than whether the water is warm or cold, but the real lesson eludes me. Be prepared to forget the thermometer left in the river, again and again. Usually you can get it the next day.

A small flashlight is essential for tying on flies in the dark and often in walking out at in the dark. Actually you can tie in the dark by sticking your fingertip in the loop of a clinch knot. Then you can stick in the loose end. Better, nobody can see you make such a crude maneuver. Add it to the list of things that you do but never talk ablut.

4.4 Entomology

This section should be called: Confessions of a Former Bug Collector. I have advised looking under rocks, not to mention shaking trees and generally keeping your eyes open, in order to find out what the bugs are doing. You can also seine for bugs in a net put in the stream. This is for real bug scientists or fanatics. Collecting insects certainly helps you catch fish. By looking under rocks you can get an idea how soon it will be before some of your favorites will hatch. However this can be taken too far, as I proved to myself a long time ago.

In my early years fishing here I went forth clanking with phials full of alcohol (the pickling kind) looking for specimens. Getting home I would spend hours trying to figure out their species or at least genus. A little of this is good in order to know a stone fly from a May fly from a caddis and some others. However too much scholarship is a mistake for various reasons. For me, to turn the fun of a fishing trip into yet another scientific investigation is perverse. (Perhaps that is because I do science for a living). You do not need to know the Latin names of all these bugs. In fact you cannot, since the diversity of species is enormous and there have not been enough entomologists available to get all these bugs under control. Just learn the basics and keep a sharp eye on the rocks. If you are going to collect insects, get pretty ones like beetles or butterflies as I did when a boy.

4.5 How to Rig and Cast Weighted Leaders

Having spent time to perfect your classic casting technique (whatever that is) you might resist a method that is awkward, prone to disastrous tanglings and physically strenuous. You have attached your 12' tapered leader to the line with a nail knot, covered with rubber cement to decrease catches in the guides. (See Appendix D for an exhortation on knots). Now take out some strips of lead and cut them into halves. Wrap the lead around the leader just above each knot, starting at the knot closest to the fly. Crimp them with your molars, to prove your virility. The number of half sections (2 centimeters in length) depends on the depth and velocity of the water. Suppose we have four weighted knots. Clearly the distribution of weight improves "casting". However some very good fishermen prefer one bigger clump of lead and perhaps also a weighted fly. However if you make a tangle through a mistake in casting, a burst of wind, or having forgotten to say your prayers to the great fish god Ichthyus, count on spending half an hour undoing the mess. Or just cut it all off and start over. After remembering to put on a fly, you are ready to begin. (Here you have a choice of weighted or unweighted flies. The unweighted fly is more buoyant but may have its nose pulled down by the lead. It swims better than the weighted nymph, which is a good snag detector).

Now for the first cast. In order to get a feel for the dynamics let the line straighten out downstream. Extend your rod backwards, farther than the usual one o'clock, and pick up quickly, anticipating the motion with a "haul" to put still more bend in the rod. When you are learning this method be sure to turn around to watch your backcast to make sure it straightens out and stays high. I often cast with my arm extended. especially for the final touch on the last forward cast. By flexing your wrist forward just as the loop goes over your head you can get still more out of the rod. It probably looks terrible.

For most purposes a motionless dead drift straight upstream is most effective. But don't forget upstream and across to get to promising lairs. Clearly the amount of lead has to be tuned to the current speed and the depth that you want. Rock bottom is not always the goal since often the trout are intercepting nymphs on their way to the surface before and even during a hatch. This situation is not always easy to solve, even if you know the menu. I usually start with a gold ribbed hare's ear, then a muskrat, caddis nymph, unless I think I know the fly, in which case an emerger pattern might be in order. In mystery cases I would use nothing bigger than size 16 to begin. As mentioned elsewhere, the wet Rio Grande King sometimes jolts fickle trout into a biting mood.

In the Fork you often need to cast across rapid currents into slower sections. This can cause fatal drag, of course. The best thing is to get close enough that you can reach over the faster current. Otherwise you will have to mend the line by flipping the developing loop downstream. It helps to throw an upstream loop in your original cast. Another sloppy solution is to throw a lot of extra line. I do this to reach the quiet water by a rock, when a rapid current lies in the way. Of course you might miss the strike that you otherwise would not have got. With the King it is all right to have drag. You can just work it as you would in a lake.

At all times you must keep the line under control by "stripping" it in. (Who is going to write the first book on the etymology of fishing?) Sometimes you have to strip rapidly. You cannot count on the fish hooking itself.

For more detail about casting and some great pictures, consult Odier's book "Swimming Flies".

4.6 Essay on Worm Fishing and Nymph Fishing

You might wonder how the author became so adept at catching trout. It is not enough to spend thousands of hours on the stream. Memorizing hatches and carrying two hundred patterns in many sizes only earns the contempt of a real fisherman.

The first secret is to learn to think like a fish. There is no substitute for beginning your angling career as a small boy. At age four I was catching them with my hands. In any case the most difficult problem for the late starter is how to tell when a trout is mouthing your nymph in relatively calm water. Let me tell you, I just know, even if there is no visible or tactile sign. When you suspect a fish presence give a little tug on the line and there is often a trout there. Most bites, not strikes, are of this kind. Unfortunately I learned this at such a young age that I cannot explain how I "know". Fishing with worms can't be beat for education. With bait-stealers all around you learn quickly. Carp fishing with dough balls on a fly rod is another great exercise. Of course sometimes they go tap-tap-tap kissing the bait just before a big run (if you think trout are strong, try a six pound carp on a fly rod. The style is more like a bulldozer than a motorcycle).

In the absence of supernatural powers allotted to the young, you can learn to anticipate the most likely part of your drift, and snug up your line. If your nymph is not inhaled as it sinks, it should be cruising near the bottom. This is a prime place for a strike. At the end of the drift the fly will speed up and turn around, giving a last chance for the greedy trout.

In fast riffles, the trout are often hooked automatically. In deep water any pause in the contour usually indicates that you have hit the bottom, but it could be a fish. The best response is a gentle pull.

In the meantime get a long cane pole, some silk black line, a few Eagle Claw hooks with an 8" leader and a coffee can of redworms driven hyperactive in coffee grounds and rotten tomatoes. Begin with a bobber to learn the strange way that fish approach the problem of fooling you.

Poem: Indiana Days

At dawn I waked to dove calls
I walked to the lake
No one was awake
Save the turtles

Then I stalked the savage fish
They dared to attack my innocent baits

All day long I lay in the boat
Rocking in the waves
Ducks flew by
My bobber detected fish hunger

All around were birds
My boat drifted into shallows
Underwater mysteries appeared

My redworms were stolen by perch
Magnificent bluegills appeared from the deep
A gift from God

Baby turtles crept from the sand
In back Inlets

That was my boyhood

Chapter 5

Some Favorite Flies

5.1 Introduction

For fishing purposes, at least, the Fork has fewer species, especially among May flies, that count, as compared with the Frying Pan. Of course the Fork has its own difficulties, and occasionally there is a selection problem that can defy solution. After the great ephemerella hatch at the end of July the trout become picky, usually in the second week of August. A common symptom is short strikes or just mouthing the fly. Besides not inhaling, some trout make an especially insulting gesture by jumping over the fly to inspect its validity from above.

In these circumstances you should experiment. Sometimes a size 16 Pink Lady will get their attention. In fact, some pink in a fly often turns defeat to triumph. Light Cahills and Ginger Quills are worth a try in small sizes. Also various terrestrials can be good. I used to think that the trout were stuffed on Green Drakes; now I think that I don't know what I am doing.

After a week of easy fishing it is possible to get bored. Then divine retribution sets in and you can't buy a bite. Or the weather turns bad. Sometimes ten days of cold and wet gray weather that put a damper on everything.

Extensive caddis hatches occur frequently. As I write (September 1) the rivers and branches are loaded with the caddis. That does not imply that the trout are not eating something else on the sly. Of course the Adams imitates many Caddis as well as other insects. In addition there are nymph versions which can save the day. Many of these, for example the muskrat nymph, are easy to tie.

It is said that the wet Rio Grand King imitates the Caddis. Beats me; I just thought that they were provocative.

I do not use midges much on the Fork. That could be a mistake. Certainly they are necessary on the Frying Pan and in mountain lakes. In addition few use streamers or bucktails. Why? I don't know, exactly. I have always liked the gray ghost streamer as a large trout getter, for example in the San Juan and Chama rivers in New Mexico. Please try midges and streamers and let me know the results.

Stoneflies usually occur in June and have disappeared before I arrive. However a few keep hatching long after this. Two inch larvae can be found under the rocks throughout the season. I need to study this situation in more detail.

5.2 Nymphs and Wet Flies

Names:	Sizes:
Gold-ribbed Hare's Ear	10-18
Caddis nymph,	10-18
(cased and uncased, dark green to bright green)	
Emergers	12-18
Muskrat	12-18
Rio Grand King, (calf tail)	8-16
Wooly worm	4-10, 2X long
Renegade	12-18
Black ant	10-16
Gray Ghost streamer	2-10
Sir Francis Drake Emerger	10-12

5.3 Dry Flies

Names:	Sizes:
Adams	12-20
Royal Wulff	6-14
Black Wulff	6-12
Sir Francis Drake, (can fish wet)	10-12, 1X long
Renegade	14-20
Pink Lady	14-16-18
Mosquito	16-20
Black Ant	12-18
Humpy	12-18
Light Cahill	12-20
Ginger Quill	12-20

Nymph fishing in complex waters like the Fork is an Aspen specialty, having been developed by George Fothergill and George Odier. Odier has written a good book on this subject.

5.4 Sir Francis Drake and Other Tying Advice

It is not our intent to write in any detail about the techniques developed for tying flies. There are many books on this subject. However I cannot resist giving a little advice. This is laid out in Appendix A, which describes the construction of some basic flies for the area.

I strongly urge you to make your own flies; you can experiment and will be more sensitive to what you are doing on the stream. If you are a beginner, pick a few basic flies and practice. Take your time and be resigned that the first ten will not be very good. In Appendix A we describe how to make eight basic, mostly easy flies.

The most important hatch is that of the Green Drake. It is a giant ephemerella, whose exact species seems uncertain, at least in the hands of amateurs. When this hatch is on, all except small trout will spurn any offering but a good imitation of this mayfly. After much effort I found a pattern that really works. My favorite version developed from Schwiebert's recipe for the "Great Red Quill" given at the end of Matching the Hatch". In honor of the great seafarer I propose the following name for this fly. It seems like a good counterpart to the Royal Wulff.

- **Sir Francis Drake**

	Schwiebert's recipe	**Pete's Variant**
Wings	wood duck flank	dark bluish dun hackle tips
Hackle	dark bluish gray dun	mixed light and dark dun
Body	rabbit fur dubbing	muskrat fur dubbing
Ribbing	claret cotton thread	pink or claret thread
Tails	dark bluish gray fibers	dark bluish gray fibers
Silk	olive	olive or brown
Hook size		10, 3X light, 1X long

It is important to use the long hook in order to end up with a fly that is the right size. These hooks are not always easy to find. The 1X long, 3X fine has been made by Orvis but I have not seen them lately. I have Tiemco 2X long and 1X fine also. This is a little heavy for Sir Francis but is perfect for the Royal Wulff. In any case you will need a long hook to imitate this fly. Do not be stingy with the double hackle.

Often the trout will be heading off the nymphs as they emerge. In this case fish the fly wet; you can also make up some emergers. Curiously drag does not seem to deter the trout in the Fork during this hatch. Weeks after the main hatch, fish will go after this fly. Often there are small and sporadic hatches throughout the summer.

Tie plenty of the Adams in size 14-16. They are not trivial to make but worth the attention. Be sure to get a top quality grizzly hackle. Do not use plastic dubbing. Remember that the best way to prepare the dubbing is to process it briefly in the coffee grinder, (after snipping the fur off the skin!). I use gray thread.

There are many variants of the Gold-ribbed hare's ear (grhe). Mine is simple: wood duck tail, rabbit mask dubbing for the body, "gold" ribbing and chin whiskers of gray rabbit fur.

Chapter 6

Simple Cooking Ideas

Naturally there are thousand of recipes for cooking trout. Here I only exhibit several which are tasty and simple to prepare.

- **Small trout up to 8 inches:**

  Clean well, scrub and leave heads on. Salt and pepper and coat with flour. To make more crust, first dip the trout in milk. Heat some butter (never margarine!) and olive oil to taste until hot but not burning. I prefer a thick aluminum pan for this purpose. Brown both sides until crisp and serve with lemon and watercress.

- **Medium size trout 8-14 inches:**

  This is an awkward size to saute. A good way to keep them moist is to wrap in foil (with a few slits for overpressure, and then to lay them on a not too hot charcoal grill.) Turn at least once. Spice as you like. Try a little paprika or New Mexico red chile. Some people like a little bacon for extra flavor.

- **Big trout, (say 14-20 in Aspen):**

  Clean well, place over foil, stuff belly with cooked wild rice preferably enhanced with Steinpilze or Chanterelle or other wild mushrooms. Salt and pepper. Slit foil after wrapping and bake at 350F.

- **All-purpose–Poached Trout:**

  Get a container of appropriate shape and size. Put trout in and cover with chicken broth (preferably home made) and dry white wine. Add scallions or leeks, salt and other spices to taste. Cover and simmer until done. They are done when the eyes pop out. Nice sauces can be invented to go with the trout, which I often prefer cold. Easy sauces are mayonnaise mixed with catchup or sour cream with dill and lemon juice.

- **Smoked Trout:**

  Buy a smoker from a regular supplier. Be careful which wood you choose. Mesquite is good.

Chapter 7

Short List of Angling Information

- **Books on Insects and Techniques:**

  Frank R. Steel, "Fly Fishing", (Crown Publishers, New York, 1949).

  Richard Talleur, "Trout Fishing for Trout; a Guide for the Adult Beginner", (Winchester Press, New York, 1974).

  Ernest Schwiebert, "Matching the Hatch" (The MacMillan Company, New York, 1955).

  George Odier, "Swimming Flies" (Stone Wall Press, Washington, 1984).

  John Gierach, "Fly Fishing Small Streams", (Stackpole Books, Harrisburg, 1989).

  Al Caucci and Bob Nastasi, "Instant Mayfly Identification Guide (Comparadun LtD, Tannersville, Pennsylvania, 1984).

  J. Michael Migel and Leonard M. Wright, Jr., "The Masters on the Nymph" (Nick Lyons Books, New York, 1979). See the contribution by Chuck Fothergill.

  Tom Rosenbauer, "Reading Trout Streams" (Nick Lyons Books, New York, 1988).

  Jack H. Dennis, Jr., "Western Trout Fly Tying Manual", (Snake River Books, Jackson Hole, 1974).

  Eric Leiser, "Fly Tying Materials" Crown Publishers, New York 1973).

- **Angling Books to Read for Fun:**

  Many fishing books have all the spontaneity of a boy scout manual. I am reminded of the old admonitions about "conservation" and the merits of the cold hip bath. You see, trout fishing easily provides cold hip baths and tedious technical commentary. Luckily, there are renegades around who like to have fun. Here are a few that come to mind.

  Sparse Gray Hackle, "Fishing Days and Angling Nights", copyright Alfred W. Miller, 1974 (Crown Publications, New York, 1974).

Norman MacLean, "A River Runs Through It", (University of Chicago Press, Chicago, 1989).

Milford "Stanley" Poltroon, "How to Fish Good", (Winchester Press, New York, 1974).

Izaac Walton, "The Compleat Angler" Hundreds of Editions.

Herter's Catalogues Waseca, Minnesota. Herter's salty commentaries on tackle and fishing often roam into global territory. I still prefer his rubber jawed hackle pliers to all the fancy ones I have accumulated. My latest copy is No. 83, August 1972-August 1973

T. Westwood and T.Satchell, Biblioteca Piscatoria, (Peyton and Co., London 1883), reprinted for Dawson's of Pall Mall, 1966 Annotated and amusing description of old angling books.

- **Other Useful Material:**

David Laing and Nicholas Lampiris, "Aspen High Country - The Geology" (Thunder River Press, Box 10935, Aspen, Colorado).

Pitkin County Open Space and Trails Map, (Pitkin County, 1992).

Len Shoemaker, "Roaring Fork Valley", (Sage Books, Denver, 1958).

Muriel Sibyl Wolfe, "Stampede to Timberline: The Ghost Towns and Mining Camps of Colorado", 2nd revised edition (Ohio University Press, Columbus, 1991).

Davic Acra, "Mushrooms Demystified", (Ten Speed Press, Berkeley, 1979).

Craighead, Craighead and Davis, "A Field Guide to Rocky Mountain Wildflowers", (Riverside Press, Cambridge, Massachusetts, 1963).

Warren Ohlrich "Trails", (Who Press, Aspen, 1988) There are many such guides describing basics trails in the area.

Chapter 8

Appendices

A. Recipes for Some Basic Flies

Although I promised not to say much about fly tying I think it is of interest for all fly fishermen to tie at least the basic ones. Few ingredients are required, except time. Now that the standard commercial flies cost $1.70 each (1993) it is worth your while to produce the basic ones. There is a freedom to deal with a snag high in a tree by just pulling it off. No more scratches climbing trees meant for 10 year olds.

In the following keep in mind the need for black waxed tying thread and fly head cement. This will do for all the flies listed below. Also hackle pliers are cheap and useful. A half hitch device is useful for finishing the head and generally keeping things under control. An inexpensive fly-tying vise will do. Scissors and a few other gadgets such as a thread holder are necessary.

In applying head cement avoid the hook eye or there will be problems. Although you can ream out the cemented eye with another hook on the stream, it is best to learn not to clog the hook eye.

1. **Wooly Worm**

   Ingredients

   Hook size 2-6, 2X long
   Grizzly hackle (large)
   Chenille (black, brown, green or orange)

   This is the easiest to tie. Get a 2X long wet fly hook, size 2-6. Take a large grizzly hackle and tie its small end on the hook near the bend.2- 6. Next, wrap the tying thread up to the front to within 3/16" of the hook eye. Now tie on some chenille, (black, green or orange) just ahead of the hackle. Wrap the chenille neatly to make the body, up to 3/16 of the hook eye, and tie down the chenille, snipping off the loose end. After wrapping the hackle forward in a snug spiral, fasten down and make a neat head for your creation.

   If you wish. put some lead wire on the hook shank before wrapping the chenille. If you are in a spendthrift mood you can put on a little red tail as the first move.

2. **Rio Grande King**

   Ingredients

   Size 10-16 wet fly hooks (e. g. Mustad 9671)
   Golden pheasant nape
   Black chenille
   White calf tail (be sure hairs are straight)
   Brown hackle

   In my experience this fly is best in the underwater version. Although dead drift upstream is very effective, you can swim the fly (for example in a beaver pond, or in deep fast runs) or dab it near the bottom in quiet pools. Tying is straightforward. First attach the golden pheasant tail and tie on the black chenille just in front of the tail. I usually use six-seven inches which is easy to manipulate. After tying a smooth body, tie it down and snip off the remainder. Now grasp some white calf tail (avoid wavy hair) with your dominant hand and select a bunch of hair. Snip it off. You might want to line up the tips a bit. Now hold it over the hook to assess the proper length. Snip off enough of the butt so that the wing extends to the bend of the hook, while the butt is tied down in front with 1/8 inch to spare for the head. Be quite firm in wrapping the hair down. Having left some of the butt unwrapped, trim the exposed part to help make a tapered head. For the hackle use brown (wrapped around) or some dense brown chinwhiskers. Often you will want to lead your leader and fish it just like a nymph.

3. **Gold Ribbed Hare's Ear Nymph**

   Ingredients

   Size 12-18 wet fly hooks
   Wood (or Mandarin) duck barred flank feathers
   Gold ribbing (woven, not flat)
   Hare's ear dubbing
   Dubbing wax
   Piece of gray rabbit fur

   The tail is wood duck. Tie in a few inches of ribbing and then put on the dubbing, usually but not always starting from the tail. To make the dubbing snip the fur off a hare's ear mask and mix it for a few seconds in the coffee grinder. (This is much better than the plastic stuff pushed by otherwise respectable dealers). Apply some wax to 2 inches of the thread adjacent to the hook and role in your fingers to make a humble yarn. Now wrap carefully towards the front and tie off, leaving room for the head. The body should bulge a bit in the middle. The ribbing is now wrapped in a uniform spiral and tied off. Build up the head a little, tie on a short bunch of rabbit fur at the chin and complete the head, finishing it with a few half hitches.

4. **Renegade**

   Ingredients

   14-20 dry fly hooks, standard length (e. g. Mustad 90840)
   Peacock herl
   Brown hackle
   White hackle

   Begin by wrapping the thread up and back to give the hook shank a little bulk. Now tie on a white hackle just where the hook straightens out. Take a double strand of Peacock herl, tie it on just ahead of the white hackle and wrap forwards, leaving room for the brown hackle and head at the front. If you wish to stabilize the delicate herl put some cement on the shank before winding the herl.

5. **Muskrat**

   Ingredients

   Size 14-18 wet fly hooks, standard length
   Muskrat fur Dubbing and wax
   Brown hackle

   No tail is necessary; add a brown one if you like. Apply muskrat dubbing to the re-waxed thread and wrap forwards. Tie on a brown hackle, complete the head, and you are done.

6. **Adams**

   Ingredients

   Size 12-18 dry fly hooks, standard length
   Grizzly hackle (good quality without web)
   Brown hackle
   Muskrat dubbing and wax

   Snip some fibers from large grizzly and brown necks, mix and tie on the tail. Next dub on some muskrat fur and wrap the body, leaving plenty of room for the double hackle to come. Find grizzly tips of the right size, snip to the right length and tie on, raising them vertically to make wings. To stabilize this tie a few turns of thread behind but snugly against the wings. Now put a grizzly and brown hackle before and after the wings. Wrap snugly, tying them down in succession. This can be a frustrating part of the Adams construction. Using a hackle guard is a useful way to keep the hackle out of the way of the wraps needed to make the thread. As a hint, begin with the larger sizes until you get a feeling for the proportions of this fly.

7. **Mosquito**

   Ingredients

   Size 12-20 dry fly hooks, standard length
   Grizzly hackle
   Moose mane, with both black and white fibers

Tail from grizzly hackle. Wrap black thread to front and back. Tie on white and a black moose mane hairs. Apply cement to shank to strengthen the moose, which you now wrap carefully forward to get alternating black and white. Tie on a grizzly hackle, finish the head and you are finished.

8. **Black Ant**

   Ingredients

   12-20 standard wet or dry hooks
   Black hackle
   Black floss or fine black wool

   There are many variants of the black ant. To begin tie on the floss and make a lump at the back to represent the abdomen. Then tie in a black hackle at midhook. Next make another lump ahead of the hackle, and tidy up the head. To imitate the flying ant, add transparent wings tied flat.White (or pale dun) can be used for this.

   To begin you don't need all the hook sizes I mentioned. Hooks are getting expensive so you may wish to cut your cost here. You can buy hooks in small quantities. If I were pushed to one hook size I would make everything in size 14, except of course the wooly worm, where I would buy size 4, 2X long.

B. Sample Fishing Log

To record your adventures, it is helpful to keep a record. In this way you can discover regularities, such as the dates of the appearance of various hatches, water condition, etc. With a copying machine you can duplicate the enclosed example (permission granted!). Or create your own suitable to your specific needs.

C. How to Tie the Basic Fishing Knots

Even for veterans, tying knots with fine leader in the gathering darkness, especially with middle aged eyes, can leave the fisherman tied in knots. The only solution is to practice. On the stream you may have to repair your leader several times in in afternoon. Always check your leader for abrasions, overhand knots which weaken the leader, and the junction of leader and fly, which can weaken as casting proceeds.

You can get along with three basic knots, all involving wrapping the leader around something. There are several other nice ones which you can learn by a winter fire. One is the Turle knot, which makes a strong loop and is easy to tie.

1. **Clinch Knot**

   This knot is intended to fasten the leader to the fly. Stick the leader tip through the eye of the fly (not always trivial) and pull about three inches through. Now wrap it around the incoming leader maybe five turns and insert in the loop formed by the eye of the hook. Holding the loose end firmly, slowly pull the knot tight. If you pull too fast the nylon will become brittle.

There is an "improved" clinch knot, formed by putting the loose end through the top loop. It is harder to snug up but stronger, especially if your leader is fine and the hook eye large. Always check the connection with a firm pull.

2. **Barrel Knot**

   This is a basic way to join two pieces of leader together. Success requires that the diameters not be too different. This knot is simply a double clinch knot. Cross the two pieces to be joined. Each piece should protrude about three inches. If you are stingy with leader and not in practice you will have trouble. Now wrap one end around the other leader and bend around to be placed in the junction formed by this action. Hold the center in place with your thumb and forefinger to leave a big enough opening to insert the other end after it has been wrapped. Pull together carefully. Snip off the leftovers. If the two ends come out on the same side you have what is known as a blood knot. Shame on you; scholars have claimed that the blood knot is knot as strong as the barrel not. If I have several to do at once, I couldn't care less about this distinction.

3. **Nail Knot**

   This is another twist knot, but too messy to be described in words. See a book with a picture. If you are patient it is not as hard as it looks. However on the stream it is very awkward to do. In a pinch I just tie a loop in the line and fasten a leader to it with a clinch knot. Usually the knot lasts several years; you can keep reusing the hard nylon butt to fasten generations of leaders. This season I lost a whole leader when the nail knot gave way. However this loaded reel had been unused for four years.

As your knots dissolve in your fingers, you will curse, or at least want to curse. Choose gentlemanly epithets, as becomes a gentleman. If you must be crude, use traditional low key excretory expletives, preferably in a foreign language.

There are other gadgets for joining the leader and line in holy matrimony. What man has joined together let no God put asunder. One such is a thin piece of wire with a small loop that protrudes, the other end being a sort of needle with barbs along its sides, inserted into the tip of the line. I never had any trouble with this device, although the transition point from wire to line has to be watched for possible fraying.

In practicing these knots use neither too thick nor too fine leader tippets. A 1X will do fine for this purpose.

D. Edible Mushrooms You Will Encounter

Here is a short list of good and easily identifiable mushrooms. As people say, be careful with identification. I consider the ones listed almost foolproof, but the penalties of mistakes can be high. For example the common Meadow Mushroom, Agaricus campestris, very similar to the grocery store mushroom, has a visual dead ringer, Agaricus Californicus, that likes to grow under trees, and makes people sick. A test is to put some lye on the cap: the offending one will turn yellow. (Among the many mushroom books available I chose to refer to the one by Acra since it has many detailed tests for edibility). Now warned, I will not include the delicious Campestris in my list. Following is a list of good mushrooms and likely places to find them on your fishing jaunts. Carry a sharp knife and many paper bags of various sizes. Do not mix mystery mushrooms with those you intend to eat.

1. **Steinpilze:** (German. I don't know a popular English name. Note that the name means stone mushroom).

   The Latin name is Boletus edulis and edulis it is. In Italian we have Porcini (little pigs) and in French Cep. The Russian name is Borovik. Unmistakable with smooth brown cap and pores (white to yellow) instead of gills. You are likely to get a lot or none at all. Outstanding sauteed, in soups, etc. It saves well, lightly sauteed, then frozen. This approach works for most mushrooms.

   To dry, slice this and lay in a sunny place on a screen or even newspapers.

   This mushroom is found associated with the blue spruce, often at the edge of a meadow. A good hint is the presence of Amanita muscaria, the "Fly agaric" (poisonous), which has a bright red cap with cottage cheese on top.

   Many specimens are wormy and have to be dissected on the spot to save the good parts.

   There are many tasty and common boletes.

   The common and similar "redcap" can be good, although one friend of mine is allergic to it.

2. **Chanterelle:** (Pfifferlinge in German. The Latin name is Cantherellus cibarius.)

   A forest full of of golden chanterelle is a joy to the gourmet. They are found at high altitude, often in little clearings in the forest. Sometimes they are on dry rocky hillsides, in which case they might be bleached by the sun and appear as small rocks.

   There are several close species, almost all edible. Consult your mushroom book. There are no worms, and the mushroom saves well. Besides being delicious sauteed in butter, the chanterelle makes outstanding sauces to go with meat. Mix cooked mushrooms with a sauce made of veloute (some flour almost browned in butter) to which you add broth (preferably homemade) and some wine. Season to taste.

3. **Honey Mushroom:** (Latin name Armillariella mellea.)

   These golden beauties grow in clumps on dead wood in the fall. Look at the picture in a book. They are very common and widespread.

   This excellent mushroom is often bypassed from unfamiliarity. Pay attention to your book and you are safe. It is really good in cream sauce. (So are others, like the Steinpilze). Saute a little onion in butter, add the mushrooms and cook slowly. Finally add some heavy cream and salt and pepper. Cook a little longer to thicken and let the flavors mix. This goes well with trout.

4. **Hedgehog Mushroom:**

   This common fungus looks strange and unappealing, but many of us like the Aspen version. Some caps can be bitter, and should be discarded.

   Cap is scaled, "teeth" replace gills. This mushroom should be cooked slowly until tender. The onion-cream approach is good in this case.

5. **Puffballs:**

   Most puffballs in the region are small and grainy. They are always safe. If you get enough fresh ones, make a soup, adding some paprika to boost the bland flavor. These belong to the genus lycoperdon.

   The puffballs are flattened to spherical, preferring roadsides and other cuts in the forest. The giant puffball (Calvatia gigantea) grows commonly to a foot in diameter. It is a favorite, either sauteed in slices or turned into soup. To make the soup cut the mushroom into pieces (cubes for the giant one) and saute in butter before combining with the other ingredients. Add milk and cream to taste, with salt and white pepper and for me, some paprika. Or you can substitute chicken broth for the milk. Corn starch (first made into a paste with some water) can be used as a thickening if desired.

   If you see any yellow drops or coloration do not eat. It will taste awful. Freezing raw specimens elicits the yellow.

   The inside should be completely smooth and white, with no worms. One possible mistake before you are familiar with the small puffballs is confusion with baby forms of the fatal amanita genus. To rule them out cut the "puffball" vertically to see if incipient signs of cap, gills and stalk are present. If so move to the next spot.

6. **Milk Caps:**

   The genus lactarius has several edible species. I find them in meadows after a rain. Among the green vegetation you will see small orange mushrooms with depressed caps. They are brittle and stain green when handled. (They exude "milk" when cut or broken.) The experts say to avoid those with yellow or purple milk.

   The texture is somewhat grainy and slow cooking is required. Most species do not stand on their own merits but they make an excellent mushroom helper when combined with other mushrooms or used in soups.

   Although I have not had any trouble with this genus I advise studying the book to understand the possibilities.

7. **Oyster Mushroom:**

   Pleurotus ostreatus is one of the great mushrooms. It emerges in fall growing on dead logs or standing dead trunks. Cottonwood is a common host. This whitish mushroom grows in stacks or rows, each member attached to the wood by a curved stem. The odor is mild and fragrant.

   The oyster mushroom is one of the few to appear in US markets. This is in contrast to Europe, where markets have many varieties. (In this case, you should know what you are doing! I remember a hallucinogeni night in Italy).

The favorite method of preparation is to dip in egg batter, bread crumbs, cook in butter till brown.

8. **Shaggy Mane:**

Coprinus comatus is very common in meadows, former pastures, road cuts and I must say, dog routes in cities. It is a member of the self- digesting class of fungi, which turn black and inky in a short time.

Having aroused your enthusiasm, I give this mushroom a very strong endorsement. My favorite cooking method is to broil them with butter. There is a slight iodine taste which I like.

As implied, one must eat them quickly.

A related species, the "Inky Cap" (Coprinus atrementarius and relatives) is edible but for many people is not, since it does not mix with alcohol. Hallucinations and other symptoms emerge, and recur if you have a glass of wine a few days later. The latter was reported by a friend who unknowingly verified the textbook lore.

The inky cap is somewhat squat, covered with yukky looking black spots, and generally has little natural appeal. It is very easy to distinguish from the distinguished-looking shaggy mane.

E. The Ideal Fisherman's Dinner

- **Soup:** Steinpilze or giant puffball, or
- **Appetizer:** Breaded and sauteed oyster mushrooms
- **Starch:** Wild rice
- **Trout:** Floured and browned assortment of small trout (say 8" or less). Cook in a mixture of butter and olive oil. Do not use margarine or other "health oils". If you do, your scales will fall out.
- **More mushrooms:** These can be added to the rice or served as a sauce. In the latter case you may want to dispense with the soup
- **Vegetable:** Your choice
- **Dessert:** Wild raspberries with vanilla ice cream

The Book of Revelations

Peter Carruthers

Aspen and Tucson

Interior of a church: A tenor stands at the podium, before a large gathering (the "chorus"). The tenor sings; a pile of Physical Reviews burns in the background.

Scene 1

- *Tenor:*

  Lord God Almighty
  We Thank Thee for Bringing
  The Ultimate Truth
  In the Form of the j Plane

  We Shall Abjure Corrupt Idols
  And Their Obscene Practices
  Field Theory Shall Die
  A Miserable Death

  Its Predictions Shall Never Be Published
  Shall Get No Grants
  As the Power of the True God
  Is Manifest in the j Plane

- *Chorus rises to respond:*

  Holy Holy Holy
  Lord God Almighty
  Forgive Us for Our Sins
  We Shall Purge Unclean Thoughts

  We Shall Not Give Tenure to Heretics
  We Shall Reject and Burn Their Works
  We Shall Denounce the Proposals
  Of Field Theory Scum

  You Have Our Word
  Lord God Almighty

Scene 2.

- *Tenor:*

Lord God Almighty
We Thank Thee for Bringing
The Ultimate Truth
In the Form of Gauge Field Theory

We Shall Abjure Corrupt Idols
And Their Obscene Practices
The j Plane Shall Die
A Miserable Death

Its Predictions Shall Never Be Published
Shall Get No Grants
As the Power of the True God
Is Manifest in Gauge Field Theory

- *Chorus rises to respond:*

Holy Holy Holy
Lord God Almighty
Forgive Us for Our Sins
We Shall Purge Unclean Thoughts

We Shall Not Give Tenure to Heretics
We Shall Reject and Burn Their Works
We Shall Denounce the Proposals
Of j Plane Scum

You Have Our Word
Lord God Almighty

Scene 3

- *Tenor:*

Lord God Almighty
We Thank Thee for Bringing
The Ultimate Truth
In the Form of the Superstring

We Shall Abjure Corrupt Idols
And Their Obscene Practices
Gauge Field Theory Shall Die
A Miserable Death

Its Predictions Shall Never Be Published
Shall Get No Grants
As the Power of the True God
Is Manifest in the Superstring

- *Chorus rises to respond:*

Holy Holy Holy
Lord God Almighty
Forgive Us for Our Sins
We Shall Purge Unclean Thoughts

We Shall Not Give Tenure to Heretics
We Shall Reject and Burn Their Works
We Shall Denounce the Proposals
Of Gauge Field Theory Scum

You Have Our Word
Lord God Almighty

Other Possible details:

- Confessions of Sinners.
- Execution of the Nonrepentant:
  - Tied in superstrings, led to the chopping block by their colleagues, with swords over their necks held by two Neanderthals labeled DOE and NSF. (Cheering of chorus at their disgrace and dismemberment. Could follow each scene.)
  - Little girl to Mother: Mommy, what happened to Daddy? Answer: Daddy was a heretic, so God had to punish him. It is good that he is gone. Now go read your Physical Review.

3 20 '92